GROUNDWATER
AND
SURFACE WATER POLLUTION

GROUNDWATER
AND
SURFACE WATER POLLUTION

Edited by
David H.F. Liu
Béla G. Lipták

Paul A. Bouis
Special Consultant

CRC Press
Taylor & Francis Group
Boca Raton London New York

CRC Press is an imprint of the
Taylor & Francis Group, an **informa** business

CRC Press
Taylor & Francis Group
6000 Broken Sound Parkway NW, Suite 300
Boca Raton, FL 33487-2742

First issued in paperback 2019

ISBN-13: 978-1-56670-511-0 (hbk)
ISBN-13: 978-0-367-39921-4 (pbk)
Library of Congress Card Number 99-055330

Library of Congress Cataloging-in-Publication Data

Groundwater and surface water pollution / edited by David H.F. Liu, Béla G. Lipták.
        p.   cm.
    Includes bibliographical references and index.
    ISBN 1-56670-511-8 (alk paper)
    1. Groundwater—Pollution. 2. Groundwater flow. 3. Groundwater—Purification.
    4. Runoff—Management. I. Liu, David H. F. II. Lipták Béla G..
TD426 .G7277 1999
628.1'68—dc21   99-055330
                                                                        CIP

Visit the Taylor & Francis Web site at
http://www.taylorandfrancis.com

and the CRC Press Web site at
http://www.crcpress.com

Preface

Dr. David H.F. Liu passed away prior to the preparation of this book.
He will be long remembered by his coworkers,
and the readers of this book will carry his memory into the 21st Century.

Engineers respond to the needs of society with technical innovations. Their tools are the basic sciences. Some engineers might end up working *on* these tools instead of working *with* them. Environmental engineers are in a privileged and challenging position, because their tools are the totality of man's scientific knowledge, and their target is nothing less than human survival through making man's peace with nature.

The Condition of the Environment

To the best of our knowledge today, life in the universe exists only in a 10-mile-thick layer on the 200-million-square-mile surface of this planet. During the 5 million years of human existence, we lived in this thin crust of earth, air, and water. Initially man relied only on inexhaustible resources. The planet appeared to be without limits and the laws of nature directed our evolution. Later we started to supplement our muscle power with exhaustible energy sources (coal, oil, uranium) and to substitute the routine functions of our brains by machines. As a result, in some respects we have "conquered nature" and today we are directing our own evolution. Today, our children grow up in man-made environments; virtual reality or cyberspace is more familiar to them than the open spaces of meadows.

While our role and power have changed, our consciousness did not. Subconsciously we still consider the planet inexhaustible and we are still incapable of thinking in timeframes which exceed a few lifetimes. These human limitations hold risks, not only for the planet, nor even for life on this planet, but for our species. Therefore, it is necessary to pay attention not only to our physical environment but also to our cultural and spiritual environment.

It is absolutely necessary to bring up a new generation which no longer shares our deeply rooted subconscious belief in continuous growth: a new generation which no longer desires the forever increasing consumption of space, raw materials, and energy.

It is also necessary to realize that, while as individuals we might not be able to think in longer terms than centuries, as a society we must. This can and must be achieved by developing rules and regulations which are appropriate to the time-frame of the processes which we control or influence. The half-life of plutonium is 24,000 years, the replacement of the water in the deep oceans takes 1000 years. For us it is difficult to be concerned about the consequences of our actions, if those consequences will take centuries or millennia to evolve. Therefore, it is essential that we develop both an educational system and a body of law which would protect our descendants from our own shortsightedness.

Protecting life on this planet will give the coming generations a unifying common purpose. The healing of environmental ills will necessitate changes in our subconscious and in our value system. Once these changes have occurred, they will not only guarantee human survival, but will also help in overcoming human divisions and thereby change human history.

The Condition of the Waters

In the natural life cycle of the water bodies (Figure 1), the sun provides the energy source for plant life (algae), which produces oxygen while converting the inorganic molecules into larger organic ones. The animal life obtains its muscle energy (heat) by consuming these molecules and by also consuming the dissolved oxygen content of the water.

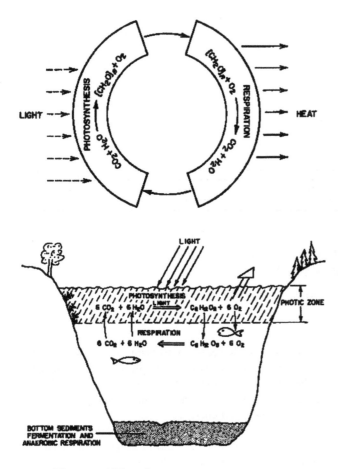

FIG. 1 The natural life cycle.

When a town or industry discharges additional organic material into the waters (which nature intended to be disposed of as fertilizer on land), the natural balance is upset. The organic effluent acts as a fertilizer, therefore the algae overpopulates and eventually blocks the transparency of the water. When the water becomes opaque, the ultraviolet rays of the sun can no longer penetrate it. This cuts off the algae from its energy source and it dies. The bacteria try to protect the life cycle in the water by attempting to break down the excess organic material (including the dead body cells of the algae), but the bacteria require oxygen for the digestion process. As the algae is no longer producing fresh oxygen, the dissolved oxygen content of the water drops, and when it reaches zero, all animals suffocate. At that point the living water body has been converted into an open sewer.

In the U.S., the setting of water quality standards and the regulation of discharges have been based on the "assimilative capacity" of the receiving waters (a kind of pollution dilution approach), which allows discharges into as yet unpolluted waterways. The Water Pollution Act of 1972 would have temporarily required industry to apply the "best practicable" and "best available" treatments of waste emissions and aimed for zero discharge by 1985. While this last goal has not been reached, the condition of

American waterways generally improved during the last decades, while on the global scale water quality has deteriorated.

Water availability has worsened since the first edition of this handbook. In the U.S. the daily withdrawal rate is about 2,000 gallons per person, which represents roughly one-third of the total daily runoff. The bulk of this water is used by agriculture and industry. The average daily water consumption per household is about 1000 gallons and, on the East Coast, the daily cost of that water is $2–$3. As some 60% of the discharged pollutants (sewage, industrial waste, fertilizers, pesticides, leachings from landfills and mines) reenter the water supplies, there is a direct relationship between the quality and cost of supply water and the degree of waste treatment in the upstream regions.

There seems to be some evidence that the residual chlorine from an upstream wastewater treatment plant can combine in the receiving waters with industrial wastes to form carcinogenic chlorinated hydrocarbons, which can enter the drinking water supplies downstream. Toxic chemicals from the water can be further concentrated through the food chain. Some believe that the gradual poisoning of the environment is responsible for cancer, AIDS, and other forms of immune deficiency and self-destructive diseases.

While the overall quality of the waterways has improved in the U.S., worldwide the opposite occurred. This is caused not only by overpopulation, but also by ocean dumping of sludge, toxins, and nuclear waste, as well as by oil leaks from off-shore oil platforms. We do not yet fully understand the likely consequences, but we can be certain that the ability of the oceans to withstand and absorb pollutants is not unlimited and, therefore, international regulations of these discharges is essential. In terms of international regulations, we are just beginning to develop the required new body of law.

Protecting the global environment and life on this planet must become a single-minded, unifying goal for all of us. The struggle will overshadow our differences, will give meaning and purpose to our lives and, if we succeed, it will mean survival for our children and the generations to come.

Béla G. Lipták

Contributors

Yong S. Chae

AB, MS, PhD, PE; Professor and Chairman,
Civil and Environmental Engineering, Rutgers University

Ahmed Hamidi

PhD, PE, PH, CGWP; Vice President,
Sadat Associates, Inc.

David H.F. Liu

PhD, ChE; Principal Scientist, J.T. Baker, Inc. a division
of Procter & Gamble

Bela G. Liptak

ME, MME, PE; Process Control and Safety Consultant,
President, Liptak Associates, P.P.

Kent Keqiang Mao

BSCE, MSCE, PhDCE, PE; President,
North America Industrial Investment Co., Ltd.

Contents

CONTENTS

Groundwater and Surface Water Pollution

Yong S. Chae | Ahmed Hamidi

2

5 Groundwater Cleanup and Remediation   85

1

Principles of Groundwater Flow

1.1
GROUNDWATER AND AQUIFERS

This section defines groundwater and aquifers and discusses the physical properties of soils, liquids, vadose zones, and aquifers.

Definition of Groundwater

Water exists in various forms in various places. Water can exist in vapor, liquid, or solid forms and exists in the atmosphere (atmospheric water), above the ground surface (surface water), and below the ground surface (subsurface water). Both surface and subsurface waters originate from precipitation, which includes all forms of moisture from clouds, including rain and snow. A portion of the precipitated liquid water runs off over the land (surface runoff), infiltrates and flows through the subsurface (subsurface flow), and eventually finds its way back to the atmosphere through evaporation from lakes, rivers, and the ocean; transpiration from trees and plants; or evapotranspiration from vegetation. This chain process is known as the hydrologic cycle. Figure 1.1.1 shows a schematic diagram of the hydrologic cycle.

Not all subsurface (underground) water is groundwater. Groundwater is that portion of subsurface water which occupies the part of the ground that is fully saturated and flows into a hole under pressure greater than atmospheric pressure. If water does not flow into a hole, where the pressure is that of the atmosphere, then the pressure in water is less than atmospheric pressure. Depths of groundwater vary greatly. Places exist where groundwater has not been reached at all (Bouwer 1978).

The zone between the ground surface and the top of groundwater is called the *vadose zone* or *zone of aeration*. This zone contains water which is held to the soil particles by capillary force and forces of cohesion and adhesion. The pressure of water in the vadose zone is negative due to the surface tension of the water, which produces a negative pressure head. Subsurface water can therefore be classified according to Table 1.1.1.

Groundwater accounts for a small portion of the world's total water, but it accounts for a major portion of the world's freshwater resources as shown in Table 1.1.2.

Table 1.1.2 illustrates that groundwater represents about 0.6% of the world's total water. However, except for glaciers and ice caps, it represents the largest source of

freshwater supply in the world's hydrologic cycle. Since much of the groundwater below a depth of 0.8 km is saline or costs too much to develop, the total volume of readily usable groundwater is about 4.2 million cubic km (Bouwer 1978).

Groundwater has been a major source of water supply throughout the ages. Today, in the United States, groundwater supplies water for about half the population and supplies about one-third of all irrigation water. Some three-fourths of the public water supply system uses groundwater, and groundwater is essentially the only water source for the roughly 35 million people with private systems (Bouwer 1978).

Aquifers

Groundwater is contained in geological formations, called *aquifers,* which are sufficiently permeable to transmit and yield water. Sands and gravels, which are found in alluvial deposits, dunes, coastal plains, and glacial deposits, are the most common aquifer materials. The more porous the material, the higher yielding it is as an aquifer material. Sandstone, limestone with solution channels, and other Karst formations are also good aquifer materials. In general, igneous and metamorphic rocks do not make good aquifers unless they are sufficiently fractured and porous.

Figure 1.1.2 schematically shows the types of aquifers. The two main types are *confined aquifers* and *unconfined aquifers.* A confined aquifer is a layer of water-bearing material overlayed by a relatively impervious material. If the confining layer is essentially impermeable, it is called an *aquiclude.* If it is permeable enough to transmit water vertically from or to the confined aquifer, but not in a horizontal direction, it is called an *aquitard.* An aquifer bound by one or two aquitards is called a *leaky* or *semiconfined aquifer.*

Confined aquifers are completely filled with groundwater under greater-than-atmospheric pressure and therefore do not have a *free water table.* The pressure condition in a confined aquifer is characterized by a *piezometric surface,* which is the surface obtained by connecting equilibrium water levels in tubes or piezometers penetrating the confined layer.

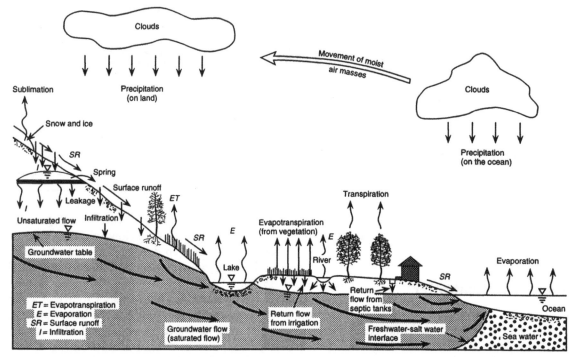

FIG. 1.1.1 Schematic diagram of the hydrologic cycle. (Reprinted from J. Bear, 1979, *Hydraulics of groundwater*, McGraw-Hill, Inc.)

An unconfined aquifer is a layer of water-bearing material without a confining layer at the top of the groundwater, called the *groundwater table*, where the pressure is equal to atmospheric pressure. The groundwater table, sometimes called the *free* or *phreatic surface*, is free to rise or fall. The groundwater table height corresponds to the equilibrium water level in a well penetrating the aquifer. Above the water table is the vadoze zone, where water pressures are less than atmospheric pressure. The soil in the vadoze zone is partially saturated, and the air is usually continuous down to the unconfined aquifer.

Physical Properties of Soils and Liquids

The following discussion describes the physical properties of soils and liquids. It also defines the terms used to describe these properties.

PHYSICAL PROPERTIES OF SOILS

Natural soils consist of solid particles, water, and air. Water and air fill the pore space between the solid grains. Soil can be classified according to the size of the particles as shown in Table 1.1.3.

Soil classification divides soils into groups and subgroups based on common engineering properties such as *texture, grain size distribution,* and Atterberg limits. The most widely accepted classification system is the unified classification system which uses group symbols for identification, e.g., SW for well-graded sand and CH for inorganic clay of high plasticity. For details, refer to any standard textbook on soil mechanics.

Figure 1.1.3 shows an element of soil, separated in three phases. The following terms describe some of the engineering and physical properties of soils used in groundwater analysis and design:

TABLE 1.1.1   CLASSIFICATION OF SUBSURFACE WATER

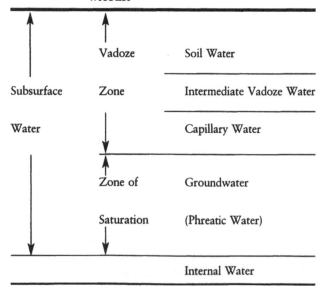

TABLE 1.1.2 ESTIMATED DISTRIBUTION OF WORLD'S WATER

	Volume 1000 km³	Percentage of Total Water
Atmospheric water	13	0.001
Surface water		
Salt water in oceans	1,320,000	97.2
Salt water in lakes and inland seas	104	0.008
Fresh water in lakes	125	0.009
Fresh water in stream channels (average)	1.25	0.0001
Fresh water in glaciers and icecaps	29,000	2.15
Water in the biomass	50	0.004
Subsurface water		
Vadose water	67	0.005
Groundwater within depth of 0.8 km	4200	0.31
Groundwater between 0.8 and 4 km depth	4200	0.31
Total (rounded)	1,360,000	100

Source: H. Bouwer, 1978, *Groundwater hydrology* (McGraw-Hill, Inc.).

POROSITY (n)—A measure of the amount of pores in the material expressed as the ratio of the volume of voids (V_v) to the total volume (V), $n = V_v/V$. For sandy soils $n = 0.3$ to 0.5; for clay $n > 0.5$.

VOID RATIO (e)—The ratio between V_v and the volume of solids V_S, $e = V_v/V_S$; where e is related to n as $e = n/(1 - n)$.

WATER CONTENT (ω)—The ratio of the amount of water in weight (W_W) to the weight of solids (W_S), $\omega = W_W/W_S$.

DEGREE OF SATURATION (S)—The ratio of the volume of water in the void space (V_W) to V_v, $S = V_W/V_v$. S varies between 0 for dry soil and 1 (100%) for saturated soil.

COEFFICIENT OF COMPRESSIBILITY (α)—The ratio of the change in soil sample height (h) or volume (V) to the change in applied pressure (σ_v)

$$\alpha = -\frac{1}{h}\frac{dh}{d\sigma_v} = -\frac{1}{V}\frac{dV}{d\sigma_v} \qquad 1.1(1)$$

The α can be expressed as

$$\alpha = \frac{(1 + \mu)(1 - 2\mu)}{E(1 - \mu)} = \frac{1}{B + \frac{4}{3}G} \qquad 1.1(2)$$

where:

E = Young's modulus
μ = Poisson's ratio
B = bulk modulus
G = shear modulus

Clay exists in either a dispersed or flocculated structure depending on the arrangement of the clay particles with

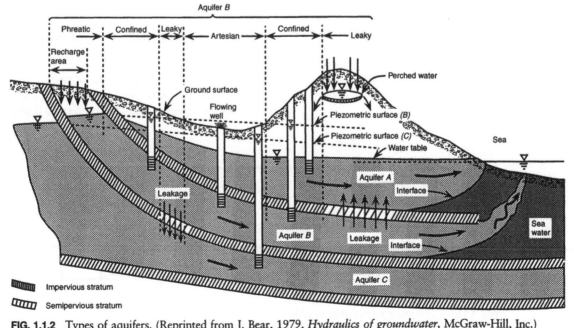

FIG. 1.1.2 Types of aquifers. (Reprinted from J. Bear, 1979, *Hydraulics of groundwater*, McGraw-Hill, Inc.)

TABLE 1.1.3   USUAL SIZE RANGE FOR GENERAL SOIL
CLASSIFICATION TERMINOLOGY

Material	Upper, mm	Lower, mm	Comments
Boulders, cobbles	1000^+	75^-	
Gravel, pebbles	75	2–5	No. 4 or larger sieve
Sand	2–5	0.074	No. 4 to No. 200 sieve
Silt	0.074–0.05	0.006	Inert
Rock flour	0.006		Inert
Clay	0.002	0.001	Particle attraction, water absorption
Colloids	0.001		

Source: J.E. Bowles, 1988, *Foundation analysis and design,* 4th ed. (McGraw-Hill).

the type of cations that are adsorbed to the clay. If the layer of adsorbed cation (such as C_a^{++}) is thin and the clay particles can be close together, making the attractive van der Waals forces dominant between the particles, then the clay is flocculated. If the clay particles are kept some distance apart by adsorbed cations (such as N_a^+), the repulsive electrostatic forces are dominant, and the clay is dispersed. Since clay particles are negatively charged, which can adsorb cations from the soil solution, clay can be converted from a dispersed state to a flocculant condition through the process of cation exchange (e.g. $N_a^+ \ominus C_a^{++}$) which changes the adsorbed ions. The reverse, changing from a flocculated to a dispersed clay, can also occur. Clay structure change is used to handle some groundwater problems in clay because the hydraulic properties of soil are dependent upon the clay structure.

PHYSICAL PROPERTIES OF WATER

The density of a material is defined as the mass per unit volume. The density (ρ) of water varies with temperature, pressure, and the concentration of dissolved materials and is about 1000 kg/m³. Multiplying ρ by the acceleration of gravity (g) gives the specific weight (γ) as $\gamma \approx \rho g$. For water, $\gamma \approx 9.8$ kN/m³.

Some of the physical properties of water are defined as follows:

DYNAMIC VISCOSITY (μ)—The ratio of shear stress (τ_{yx}) in x direction, acting on an x–y plane to velocity gradient (dv_x/dy); $\tau_{yx} = \mu \, dv_x/dy$. For water, $\mu = 10^{-3}$ kg/m · s.

KINEMATIC VISCOSITY (v)—Related to μ by $v = \mu/\rho$. Its value is about 10^{-6} m²/s for water.

COMPRESSIBILITY (β)—The ratio of change in density caused by change in pressure to the original density

$$\beta = \frac{1}{\rho} \frac{d\rho}{dp} = -\frac{1}{V} \frac{dV}{dp}$$

$$\beta \approx 0.5 \times 10^{-9} \text{ m}^2/\text{N} \qquad \text{1.1(3)}$$

The variation of density and viscosity of water with temperature can be obtained from Table 1.1.4.

Physical Properties of Vadose Zones and Aquifers

A description of the physical properties of vadose zones and aquifers follows.

PHYSICAL PROPERTIES OF VADOSE ZONES

As discussed earlier, the pressure of water in the vadose zone is negative, and the negative pressure head or capillary pressure is proportional to the vertical distance above

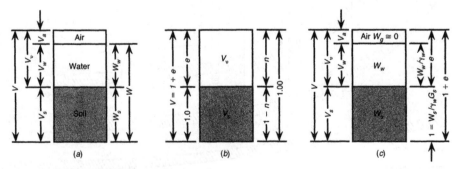

FIG. 1.1.3   Three-phase relationship in soils.

TABLE 1.1.4   VARIATION OF DENSITY AND VISCOSITY OF WATER WITH TEMPERATURE

Temperature (°C)	Density (kg/m³)	Dynamic Viscosity (kg/m s)
0	999.868	1.79×10^{-3}
5	999.992	1.52×10^{-3}
10	999.727	1.31×10^{-3}
15	999.126	1.14×10^{-3}
20	998.230	1.01×10^{-3}

Source: A. Verrjuitt, 1982, *Theory of groundwater flow*, 2d ed. (Macmillan Publishing Co.).

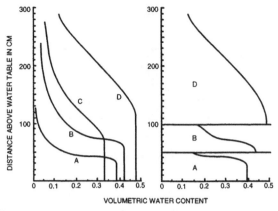

FIG. 1.1.4 Schematic equilibrium water-content distribution above a water table (left) for a coarse uniform sand (A), a fine uniform sand (B), a well-graded fine sand (C), and a clay soil (D). The right plot shows the corresponding equilibrium water-content distribution in a soil profile consisting of layers of materials A, B, and D. (Reprinted from H. Bouwer, 1978, *Groundwater hydrology*, McGraw-Hill, Inc.)

the water table. Figure 1.1.4 shows a characteristic curve of the relationship between volumetric water content and the negative pressure head (height above the water table or capillary pressure).

For materials with relatively uniform particle size and large pores, the water content decreases abruptly once the air-entry value is reached. These materials have a well-defined capillary fringe. For well-graded materials and materials with fine pores, the water content decreases more gradually and has a less well-defined capillary fringe.

At a large capillary pressure, the volumetric water content tends towards a constant value because the forces of adhesion and cohesion approach zero. The volumetric water content at this state is equal to the *specific retention*. The specific retention is then the amount of water retained against the force of gravity compared to the total volume of the soil when the water from the pore spaces of an unconfined aquifer is drained and the groundwater table is lowered.

PHYSICAL PROPERTIES OF AQUIFERS

As stated before, an aquifer serves as an underground storage reservoir for water. It also acts as a conduit through which water is transmitted and flows from a higher level to a lower level of energy. An aquifer is characterized by the three physical properties: *hydraulic conductivity, transmissivity, and storativity.*

Hydraulic Conductivity

Hydraulic conductivity, analogous to electric or thermal conductivity, is a physical measure of how readily an aquifer material (soil) transmits water through it. Mathematically, it is the proportionality between the rate of flow and the energy gradient causing that flow as expressed in the following equation. Therefore, it depends on the properties of the aquifer material (porous medium) and the fluid flowing through it.

$$K = k \frac{\gamma}{\mu} \qquad 1.1(4)$$

where:

K = hydraulic conductivity (called the coefficient of permeability in soil mechanics)
k = intrinsic permeability
γ = specific weight of fluid
μ = dynamic viscosity of fluid

For a given fluid under a constant temperature and pressure, the hydraulic conductivity is a function of the properties of the aquifer material, that is, how permeable the soil is. The subject of hydraulic conductivity is discussed in more detail in Section 1.2.

Transmissivity

Transmissivity is the physical measure of the ability of an aquifer of a known dimension to transmit water through it. In an aquifer of uniform thickness d, the transmissivity T is expressed as

$$T = \bar{K}d \qquad 1.1(5)$$

where $\bar{K}$ represents an average hydraulic conductivity. When the hydraulic conductivity is a continuous function of depth

$$\bar{K} = \frac{1}{d} \int_o^d Kz \, dz \qquad 1.1(6)$$

When a medium is stratified, either in horizontal (x) or vertical (y) direction with respect to hydraulic conductivity as shown in Figure 1.1.5, the average value $\bar{K}$ can be obtained by

$$\bar{K}_x = \sum_{m=1}^{n} \frac{K_m d_m}{d} \qquad 1.1(7)$$

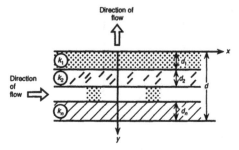

FIG. 1.1.5 Permeability of layered soils.

$$\bar{K}_y = \frac{d}{\displaystyle\sum_{m=1}^{n} \frac{d_m}{K_m}} \qquad 1.1(8)$$

Storativity

Storativity, also known as the *coefficient of storage* or *specific yield*, is the volume of water yielded or released per unit horizontal area per unit drop of the water table in an unconfined aquifer or per unit drop of the piezometric surface in a confined aquifer. Storativity S is expressed as

$$S = \frac{1}{A}\frac{dQ}{d\phi} \qquad 1.1(9)$$

where:

dQ = volume of water released or restored
$d\phi$ = change of water table or piezometric surface

Thus, if an unconfined aquifer releases 2 m³ water as a result of dropping the water table by 2m over a horizontal area of 10 m², the storativity is 0.1 or 10%.

—Y.S. Chae

Reference

Bouwer, H. 1978. *Groundwater hydrology.* McGraw-Hill, Inc.

1.2
FUNDAMENTAL EQUATIONS OF GROUNDWATER FLOW

The flow of water through a body of soil is a complex phenomenon. A body of soil constitutes, as described in Section 1.1, a solid matrix and pores. For simplicity, assume that all pores are interconnected and the soil body has a uniform distribution of phases throughout. To find the law governing groundwater flow, the phenomenon is described in terms of average velocities, average flow paths, average flow discharge, and pressure distribution across a given area of soil.

The theory of groundwater flow originates with Henry Darcy who published the results of his experimental work in 1856. He performed a series of experiments of the type shown in Figure 1.2.1. He found that the total discharge Q was proportional to cross-sectional area A, inversely proportional to the length Δs, and proportional to the head difference $\phi_1 - \phi_2$ as expressed mathematically in the form

$$Q = KA\frac{\phi_1 - \phi_2}{\Delta s} \qquad 1.2(1)$$

where K is the proportionality constant representing hydraulic conductivity. This equation is known as Darcy's equation. The quantity Q/A is called *specific discharge* q. If $\phi_1 - \phi_2 = \Delta\phi$ and $\Delta s \longrightarrow 0$, Equation 1.2(1) becomes

$$q = -K\frac{d\phi}{ds} \qquad 1.2(2)$$

This equation states that the specific discharge is directly proportional to the derivative of the head in the direction of flow (hydraulic gradient). The specific discharge is also known as Darcy's velocity. Note that q is not the actual flow velocity (seepage velocity) because the flow is limited to pore space only. The seepage velocity v is then

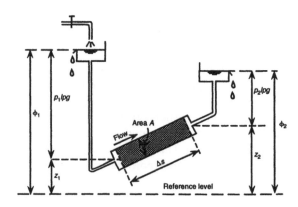

FIG. 1.2.1 Darcy's experiment.

$$v = \frac{Q}{n \cdot A} = \frac{q}{n} \qquad \text{1.2(3)}$$

where n is the porosity of the soil. Note that v is always larger than q.

Intrinsic Permeability

The hydraulic conductivity K is a material constant, and it depends not only on the type of soil but also on the type of fluid (dynamic viscosity μ) percolating through it. The hydraulic conductivity K is expressed as

$$K = k \frac{\gamma}{\mu} \qquad \text{1.2(4)}$$

where k is called the intrinsic permeability and is now a property of the soil only. Many attempts have been made to express k by such parameters as average pore diameter, porosity, and effective soil grain size. The most familiar equation is that of Kozeny-Carmen

$$k = Cd^2 \frac{n^3}{(1 - n)^2} \qquad \text{1.2(5)}$$

where:

n = porosity
d = the effective pore diameter
C = a constant to account for irregularities in the geometry of pore space

Another equation by Hazen states

$$k = CD^2 = C_1 D_{10}^2 \qquad \text{1.2(6)}$$

where:

D   = the average grain diameter
D_{10} = the effective diameter of the grains retained

Values of hydraulic conductivity can be obtained from empirical formulas, laboratory experiments, or field tests. Table 1.2.1 gives the typical values for various aquifer materials.

Validity of Darcy's Law

Darcy's law is restricted to a specific discharge less than a certain critical value and is valid only within a laminar

TABLE 1.2.1   THE ORDER OF MAGNITUDE OF THE PERMEABILITY OF NATURAL SOILS

	k (m²)	K (m/s)
Clay	10^{-17} to 10^{-15}	10^{-10} to 10^{-8}
Silt	10^{-15} to 10^{-13}	10^{-8} to 10^{-6}
Sand	10^{-12} to 10^{-10}	10^{-5} to 10^{-3}
Gravel	10^{-9} to 10^{-8}	10^{-2} to 10^{-1}

Source: A. Verrjuit, 1982, *Theory of groundwater flow*, 2d ed. (Macmillan Publishing Co.).

flow condition, which is expressed by Reynolds number R_e defined as

$$R_e = \frac{qD\rho}{\mu} = \frac{qD}{\nu} \qquad \text{1.2(7)}$$

Experiments have shown the range of validity of Darcy's law to be

$$R_e \leq 1 \sim 10 \qquad \text{1.2(8)}$$

In practice, the specific discharge is always small enough for Darcy's law to be applicable. Only cases of flow through coarse materials, such as gravel, deviate from Darcy's law. Darcy's law is not valid for flow through extremely fine-grained soils, such as colloidal clays.

Generalization of Darcy's Law

In practice, flow is seldom one dimensional, and the magnitude of the hydraulic gradient is usually unknown. The simple form, Equation 1.2(2), of Darcy's law is not suitable for solving problems. A generalized form must be used, assuming the hydraulic conductivity K to be the same in all directions, as

$$q_x = -K \frac{\partial \phi}{\partial x}$$

$$q_y = -K \frac{\partial \phi}{\partial y}$$

$$q_z = -K \frac{\partial \phi}{\partial z} \qquad \text{1.2(9)}$$

For an anisotropic material, these equations can be written as

$$q_x = -K_{xx} \frac{\partial \phi}{\partial x} - K_{xy} \frac{\partial \phi}{\partial y} - K_{xz} \frac{\partial \phi}{\partial z}$$

$$q_y = -K_{yx} \frac{\partial \phi}{\partial x} - K_{yy} \frac{\partial \phi}{\partial y} - K_{yz} \frac{\partial \phi}{\partial z}$$

$$q_z = -K_{zx} \frac{\partial \phi}{\partial x} - K_{zy} \frac{\partial \phi}{\partial y} - K_{zz} \frac{\partial \phi}{\partial z} \qquad \text{1.2(10)}$$

In the special case that $K_{xy} = K_{xz} = K_{yx} = K_{yz} = K_{zx} = K_{zy} = 0$, the x, y, and z directions are the principal directions of permeability, and Equations 1.2(10) reduce to

$$q_x = -K_{xx} \frac{\partial \phi}{\partial x} = -K_x \frac{\partial \phi}{\partial x}$$

$$q_y = -K_{yy} \frac{\partial \phi}{\partial y} = -K_y \frac{\partial \phi}{\partial y}$$

$$q_z = -K_{zz} \frac{\partial \phi}{\partial z} = -K_z \frac{\partial \phi}{\partial z} \qquad \text{1.2(11)}$$

This chapter considers isotropic soils since problems for anisotropic soils can be easily transformed into problems for isotropic soils.

Equation of Continuity

Darcy's law furnishes three equations of motion for four unknowns (q_x, q_y, q_z, and ϕ). A fourth equation notes that the flow phenomenon must satisfy the fundamental physical principle of conservation of mass. When an elementary block of soil is filled with water, as shown in Figure 1.2.2, no mass can be gained or lost regardless of the pattern of flow.

The conservation principal requires that the sum of the three quantities (the mass flow) is zero, hence when divided by $\Delta x \cdot \Delta y \cdot \Delta z$

$$\frac{\partial(\rho q_x)}{\partial x} + \frac{\partial(\rho q_y)}{\partial y} + \frac{\partial(\rho q_z)}{\partial z} = 0 \qquad 1.2(12)$$

When the density is a constant, then Equation 1.2(12) is reduced to

$$\frac{\partial q_x}{\partial x} + \frac{\partial q_y}{\partial y} + \frac{\partial q_z}{\partial z} = 0 \qquad 1.2(13)$$

This equation is called the equation of continuity.

Fundamental Equations

Darcy's law and the continuity equation provide four equations for the four unknowns. Substituting Darcy's law Equation 1.2(9) into the equation of continuity Equation 1.2(13) yields

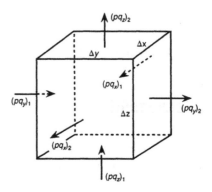

FIG. 1.2.2 Conservation of mass.

$$\frac{\partial^2 \phi}{\partial x^2} + \frac{\partial^2 \phi}{\partial y^2} + \frac{\partial^2 \phi}{\partial z^2} = 0 \qquad 1.2(14)$$

or

$$\nabla^2 \phi = 0 \qquad 1.2(15)$$

which is Laplace's equation in three dimensions.

Solving groundwater flow problems amounts to solving Laplace's equation with the appropriate boundary conditions. It is essentially a mathematical problem. Sometimes a problem must be simplified before it can be solved, and these simplifications involve considering the physical condition of groundwater flow.

—*Y.S. Chae*

1.3
CONFINED AQUIFERS

This section discusses groundwater flow in confined aquifers including one-dimensional horizontal flow, semiconfined flow, and radial flow. It also discusses radial flow in a semiconfined aquifer.

One-Dimensional Horizontal Flow

One-dimensional horizontal confined flow means that water is flowing through a confined aquifer in one direction only. Figure 1.3.1 shows an example of such a flow. Since $q_y = q_z = 0$, the governing Equation 1.2(14) reduces to

$$\frac{d^2 \phi}{dx^2} = 0 \qquad 1.3(1)$$

and the general solution of this equation is $\phi = Ax + B$. Using the boundary conditions from Figure 1.3.1 of

$x = 0 \qquad \phi = \phi_1$

$x = L \qquad \phi = \phi_2$

gives

$$\phi = \phi_1 - \frac{\phi_1 - \phi_2}{L} x \qquad 1.3(2)$$

Equation 1.3(2) indicates that the piezometric head ϕ decreases linearly with distance. The specific discharge q_x is

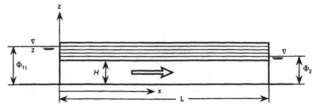

FIG. 1.3.1 One-dimensional flow in a confined aquifer.

then found using Darcy's law

$$q_x = -K \frac{\partial \phi}{\partial x} = K \frac{\phi_1 - \phi_2}{L} \qquad 1.3(3)$$

which follows that the specific discharge does not vary with position. The discharge flowing through the aquifer Q_x per unit length of the river bank is then

$$Q_x = q_x \cdot H = KH \frac{\phi_1 - \phi_2}{L} \qquad 1.3(4)$$

Semiconfined Flow

If an aquifer is bound by one or two aquitards which allow water to be transmitted vertically from or to the confined aquifer as shown in Figure 1.3.2, then a semiconfined or leaky aquifer exists, and the flow through this aquifer is called *semiconfined flow*. Small amounts of water can enter (or leave) the aquifer through the aquitards of low permeability, which cannot be ignored. Yet in the aquifer proper, the horizontal flow dominates ($q_z = o$ is assumed).

The fundamental equation of semiconfined flow is derived from the principle of continuity and Darcy's law as follows:

Consider an element of the aquifer shown in Figure 1.3.2. The net outward flux due to the flow in x and y directions is

$$-K\left(\frac{\partial^2 \phi}{\partial x^2} + \frac{\partial^2 \phi}{\partial y^2}\right) \Delta x \cdot \Delta y \cdot H \qquad 1.3(5)$$

The amount of water percolating through the layers per unit time is

$$K_1 \frac{\phi - \phi_1}{d_1} \Delta x \cdot \Delta y$$

$$K_2 \frac{\phi - \phi_2}{d_2} \Delta x \cdot \Delta y \qquad 1.3(6)$$

Continuity now requires that the sum of these quantities be zero, hence

$$KH\left(\frac{\partial^2 \phi}{\partial x^2} + \frac{\partial^2 \phi}{\partial y^2}\right) - \frac{\phi - \phi_1}{c_1} - \frac{\phi - \phi_2}{c_2} = 0 \qquad 1.3(7)$$

where $c_1 = d_1/K_1$ and $c_2 = d_2/K_2$, which are called *hydraulic resistances* of the confining layers. The terms $(\phi - \phi_1)/c_1$ and $(\phi - \phi_2)/c_2$ represent the vertical leakage through the confining layers.

Defining *leakage factor* $\lambda = \sqrt{Tc}$ where $T = KH$, the transmissivity of the aquifer, Equation 1.3(7), can be written as

$$\frac{\partial^2 \phi}{\partial x^2} + \frac{\partial^2 \phi}{\partial y^2} - \frac{\phi - \phi_1}{\lambda^2_1} - \frac{\phi - \phi_2}{\lambda^2_2} = 0 \qquad 1.3(8)$$

This equation is the fundamental equation of semiconfined flow. When the confining layers are completely impermeable ($K_1 = K_2 = 0$), Equation 1.3(8) reduces to Equation 1.2(14).

Radial Flow

Radial flow in a confined aquifer occurs when the flow is symmetrical about a vertical axis. An example of radial flow is that of water pumped through a well in an open field or a well located at the center of an island as shown in Figure 1.3.3. The distance R, called the *radius of influence zone,* is the distance to the source of water where the piezometric head ϕ_0 does not vary regardless of the amount of pumping. The radius R is well defined in the case of pumping in a circular island. In an open field, however, the distance R is theoretically infinite, and a steady-state solution cannot be obtained. In practice, this case does not occur, and R can be obtained by empirical formula or measurements.

The differential equation governing radial flow is obtained when the cartesian coordinates used for rectilinear flow are transformed into polar coordinates as

$$\frac{\partial^2 \phi}{\partial x^2} + \frac{\partial^2 \phi}{\partial y^2} = \frac{\partial^2 \phi}{\partial r^2} + \frac{1}{r}\frac{\partial \phi}{\partial r} + \frac{1}{r^2}\frac{\partial^2 \phi}{\partial r^2} + \frac{1}{r^2}\frac{\partial^2 \phi}{\partial \theta^2} \qquad 1.3(9)$$

Since ϕ is independent of angle θ, the last term of this equation can be dropped. The fundamental equation of

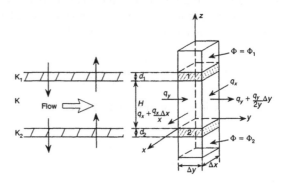

FIG. 1.3.2 Semiconfined flow.

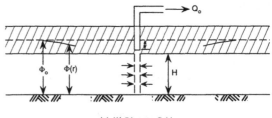

(a) Well in open field

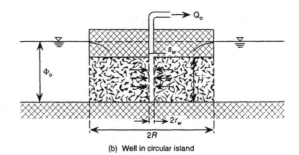

(b) Well in circular island

FIG. 1.3.3 Radial flow in a confined aquifer.

radial flow is then

$$\frac{\partial^2 \phi}{\partial r^2} + \frac{1}{r}\frac{\partial \phi}{\partial r} = 0 \qquad \text{1.3(10)}$$

or

$$\frac{1}{r}\frac{d}{dr}\left(r\frac{d\phi}{dr}\right) = 0 \qquad \text{1.3(11)}$$

The solution of this differential equation with boundary conditions (Gupta 1989) yields

$$\phi = \frac{Q}{2\pi KH}\ln\frac{r}{R} + \phi_o \qquad \text{1.3(12)}$$

This equation is known as the Thiem equation.

To calculate the head at the well ϕ_w using Equation 1.3(12), substitute the radius of the well r_w for r, which gives

$$\phi_w = \frac{Q}{2\pi KH}\ln\left(\frac{r_w}{R}\right) + \phi_o \qquad \text{1.3(13)}$$

Since the flow is confined, the head at the well must be above the upper impervious boundary (ϕ must be greater than H). Otherwise, the flow in that situation becomes unconfined flow, and Equation 1.3(13) is not applicable.

If the radius of influence zone is known or can be determined, the discharge rate is obtained by

$$Q_o = 2\pi KH \frac{\phi_o - \phi_w}{\ln\left(\dfrac{R}{r_w}\right)} \qquad \text{1.3(14)}$$

and the drawdown s at any point is given by

$$s = \phi_o - \phi = \frac{Q}{2\pi KH}\ln\left(\frac{R}{r}\right) \qquad \text{1.3(15)}$$

Radial Flow in a Semiconfined Aquifer

Radial flow in a semiconfined aquifer occurs when the flow is towards a well in an aquifer such as the one shown in Figure 1.3.4.

When leakage through the confining layer is considered, Equation 1.3(4) becomes

$$\frac{\partial^2 \phi}{\partial r^2} + \frac{1}{r}\frac{\partial \phi}{\partial r} - \frac{\phi - \phi_1}{\lambda^2_1} = 0 \qquad \text{1.3(16)}$$

The general solution of this equation is

$$\phi = \phi_o + AI_o\left(\frac{r}{\lambda}\right) + BK_o\left(\frac{r}{\lambda}\right) \qquad \text{1.3(17)}$$

where A and B are arbitrary constants, and I_o and K_o are modified Bessel functions of zero order and of the first and second kind, respectively. Table 1.3.1 is a short table of the four types of Bessel functions. The two constants are determined with the two boundary conditions as $r \ominus \infty$, $\phi = \phi_o$ and $r - r_w$, $Q_o = -2\pi r Hq_r$. The solution of this equation is then

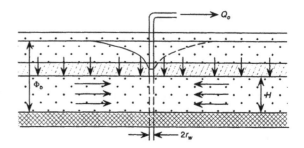

FIG. 1.3.4 Radial flow in an infinite semiconfined aquifer. (Reprinted from A. Verrjuit, 1982, *Theory of groundwater flow*, 2d ed., Macmillan Pub. Co.)

TABLE 1.3.1   BESSEL FUNCTIONS

x	$I_0(x)$	$I_1(x)$	$K_0(x)$	$K_1(x)$
0.0	1.0000	0.0000	∞	∞
0.1	1.0025	0.0501	2.4271	9.8538
0.2	1.0100	0.1005	1.7527	4.7760
0.3	1.0226	0.1517	1.3725	3.0560
0.4	1.0404	0.2040	1.1145	2.1844
0.5	1.0635	0.2579	0.9244	1.6564
0.6	1.0920	0.3137	0.7775	1.3028
0.7	1.1263	0.3719	0.6605	1.0503
0.8	1.1665	0.4329	0.5653	0.8618
0.9	1.2130	0.4971	0.4867	0.7165
1.0	1.2661	0.5652	0.4210	0.6019
1.1	1.3262	0.6375	0.3656	0.5098
1.2	1.3937	0.7147	0.3185	0.4346
1.3	1.4693	0.7973	0.2782	0.3726
1.4	1.5534	0.8861	0.2436	0.3208
1.5	1.6467	0.9817	0.2138	0.2774
1.6	1.7500	1.0848	0.1880	0.2406
1.7	1.8640	1.1963	0.1655	0.2094
1.8	1.9896	1.3172	0.1459	0.1826
1.9	2.1277	1.4482	0.1288	0.1597
2.0	2.2796	1.5906	0.1139	0.1399
2.1	2.4463	1.7455	0.1008	0.1228
2.2	2.6291	1.8280	0.0893	0.1079
2.3	2.8296	2.0978	0.0791	0.0950
2.4	3.0493	2.2981	0.0702	0.0837
2.5	3.2898	2.5167	0.0624	0.0739
2.6	3.5533	2.7554	0.0554	0.0653
2.7	3.8416	3.0161	0.0493	0.0577
2.8	4.1573	3.3011	0.0438	0.0511
2.9	4.5028	3.6126	0.0390	0.0453
3.0	4.8808	3.9534	0.0347	0.0402
3.1	5.2945	4.3262	0.0310	0.0356
3.2	5.7472	4.7342	0.0276	0.0316
3.3	6.2426	5.1810	0.0246	0.0281
3.4	6.7848	5.6701	0.0220	0.0250
3.5	7.3782	6.2058	0.0196	0.0222
3.6	8.0277	6.7927	0.0175	0.0198
3.7	8.7386	7.4358	0.0156	0.0176
3.8	9.5169	8.1404	0.0140	0.0157
3.9	10.3690	8.9128	0.0125	0.0140
4.0	11.3019	9.7595	0.0112	0.0125

$$\phi = \phi_o - \frac{Q_o}{2\pi T} K_o\left(\frac{r}{\lambda}\right) \qquad 1.3(18)$$

When r approaches 4λ, K_o (4) approaches zero which means that at $r > 4\lambda$, drawdown is practically negligible. Note that when $r/\lambda << 1$, $K_o(r/\lambda) \approx -\ln(r/1.123\lambda)$, ϕ becomes

$$\phi = \phi_o + \frac{Q_o}{2\pi T} \ln\left(\frac{r}{1.123\lambda}\right) \qquad 1.3(19)$$

This equation is similar to the governing equation for a confined aquifer, Equation 1.3(13), with the equivalent radius R_{eq} equal to 1.123λ. Therefore, the equation can be rewritten as

$$\phi = \phi_o + \frac{Q_o}{2\pi T} \ln\left(\frac{r}{R_{eq}}\right) \qquad 1.3(20)$$

Equation 1.3(20) indicates that the drawdown near the well s_w can be expressed as

$$s_w = \phi_o - \phi_w = -\frac{Q_o}{2\pi T} \ln\left(\frac{r}{1.123\lambda}\right) \qquad 1.3(21)$$

Basic Equations

The fundamental equations of groundwater flow can be derived in terms of the discharge vector Q_i rather than the specific discharge q_i. For two-dimensional flow, the discharge vector has two components Q_x and Q_y and is defined as

$$Q_x = Hq_x$$
$$Q_y = Hq_y \qquad 1.3(22)$$

With the use of Darcy's law

$$Q_x = Hq_x = H\left(-K\frac{\partial \phi}{\partial x}\right)$$
$$Q_y = Hq_y = H\left(-K\frac{\partial \phi}{\partial y}\right) \qquad 1.3(23)$$

These equations can be rewritten as

$$Q_x = -\frac{\partial(KH\phi)}{\partial x}$$
$$Q_y = -\frac{\partial(KH\phi)}{\partial y} \qquad 1.3(24)$$

With the substitution of a new variable Φ, defined as

$$\Phi = KH\phi + C_c \qquad 1.3(25)$$

where C_c is an arbitrary constant, Equations 1.3(24) can be simplified since the derivatives of C_c with respect x and y are zero as

$$Q_x = -\frac{\partial \Phi}{\partial x}$$
$$Q_x = -\frac{\partial \Phi}{\partial y} \qquad 1.3(26)$$

The function Φ is referred to as the *discharge potential* for horizontal flow or simply as the *potential*.

Now the governing equation for horizontal confined flow, Equation 1.2(13), expressed in terms of the head ϕ is

$$\frac{\partial^2 \phi}{\partial x^2} + \frac{\partial^2 \phi}{\partial y^2} = 0 \qquad 1.3(27)$$

and can be written in terms of the potential Φ as

$$\frac{\partial^2 \Phi}{\partial x^2} + \frac{\partial^2 \phi}{\partial y^2} = 0 \qquad 1.3(28)$$

or

$$\nabla^2 \Phi = 0 \qquad 1.3(29)$$

Solutions to horizontal confined flow can be obtained when Φ is determined from this Laplace's equation with proper boundary conditions satisfied.

The following equations give solutions for horizontal confined flow in terms of Φ.

(1) One-dimensional flow

$$\Phi = KH\phi = \Phi_1 - \frac{\Phi_1 - \Phi_2}{L} x \qquad 1.3(30)$$

(2) Radial flow

$$\Phi = KH\phi = \frac{Q}{2\pi} \ln\frac{r}{R} + \Phi_o \qquad 1.3(31)$$

Two-dimensional flow problems expressed by the differential Equation 1.3(29) are discussed in more detail in Section 2.1.

—Y.S. Chae

Reference

Gupta, R.S. 1989. *Hydrology and hydraulic systems*. Prentice-Hall, Inc.

1.4
UNCONFINED AQUIFERS

As defined in Section 1.1, an unconfined aquifer is a water-bearing layer whose upper boundary is exposed to the open air (atmospheric pressure), as shown in Figure 1.4.1, known as the phreatic surface. Problems with such a boundary condition are difficult to solve, and the vertical component of flow is often neglected. The Dupuit-Forchheimer assumption to neglect the variation of the piezometric head with depth ($\partial\phi/\partial z = 0$) means that the head along any vertical line is constant ($\phi = h$). Physically, this assumption is not true, of course, but the slope of the phreatic surface is usually small so that the variation of the head horizontally ($\partial\phi/\partial x$, $\partial\phi/\partial y$) is much greater than the vertical value of $\partial\phi/\partial z$. The basic differential equation for the flow of groundwater in an unconfined aquifer can be derived from Darcy's law and the continuity equation.

Discharge Potential and Continuity Equation

The discharge vector, as defined in Section 1.3, is the product of the specific discharge q and the thickness of the aquifer H. For an unconfined aquifer, the aquifer thickness h varies, and thus

$$Q_x = q_x h = -Kh\frac{\partial\phi}{\partial x}$$

$$Q_y = q_y h = -Kh\frac{\partial\phi}{\partial y} \qquad \textbf{1.4(1)}$$

Since $h = \phi$ and K is a constant, Equation 1.4(1) becomes

$$Q_x = -\frac{\partial}{\partial x}\left(\frac{1}{2}K\phi^2\right)$$

$$Q_y = -\frac{\partial}{\partial y}\left(\frac{1}{2}K\phi^2\right) \qquad \textbf{1.4(2)}$$

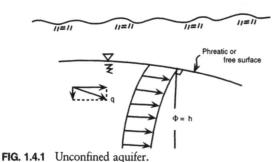

FIG. 1.4.1 Unconfined aquifer.

the discharge potential for unconfined flow introducing as

$$\Phi = \frac{1}{2}K\phi^2 + C_u \qquad \textbf{1.4(3)}$$

where C_u is an arbitrary constant. Now Equations 1.4(2) can be rewritten as

$$Q_x = -\frac{\partial\Phi}{\partial x}$$

$$Q_y = -\frac{\partial\Phi}{\partial y} \qquad \textbf{1.4(4)}$$

These equations are the same as those derived for confined flow, Equation 1.3(26).

The continuity equation for unconfined flow, without regard for inflow or outflow along the upper boundary due to precipitation or evaporation, is the same as that for confined flow as

$$\frac{\partial Q_x}{\partial x} + \frac{\partial Q_y}{\partial y} = 0 \qquad \textbf{1.4(5)}$$

Basic Differential Equation

The governing equation for unconfined flow is obtained when Equation 1.4(4) is substituted into Equation 1.4(5) as

$$\frac{\partial^2\Phi}{\partial x^2} + \frac{\partial^2\Phi}{\partial y^2} = 0 \qquad \textbf{1.4(6)}$$

The governing equation for both confined and unconfined flows is the same, in terms of the discharge potential, and problems can be solved in the same manner mathematically. The only difference between confined and unconfined flows lies in the expression for Φ as

$$\Phi = KH\phi + C_c \quad \text{for confined flow} \qquad \textbf{1.4(7)}$$

and

$$\Phi = \frac{1}{2}K\phi^2 + C_u \quad \text{for unconfined flow} \qquad \textbf{1.4(8)}$$

One-Dimensional Flow

The simplest example of unconfined flow is that of an unconfined aquifer between two long parallel bodies of water, such as rivers or canals, as shown in Figure 1.4.2. In this case, ϕ is a function of x only, and the differential Equation 1.4(6) reduces to

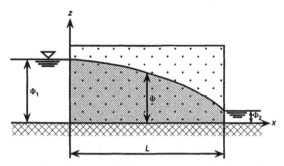

FIG. 1.4.2 One-dimensional flow in an unconfined aquifer.

$$\frac{d^2\Phi}{dx^2} = 0 \qquad 1.4(9)$$

with the general solution

$$\Phi = Ax + B \qquad 1.4(10)$$

Constants A and B can be found from the boundary conditions

$$x = 0, \quad \Phi = \Phi_1 \quad B = \Phi_1$$

$$x = L, \quad \Phi = \Phi_2 \quad A = \frac{\Phi_2 - \Phi_1}{L}$$

Substitution of A and B into Equation 1.4(10) yields

$$\Phi = \frac{\Phi_2 - \Phi_1}{L}x + \Phi_1 \qquad 1.4(11)$$

An expression for the head ϕ can be found by

$$\frac{1}{2}K\phi^2 = \frac{\frac{1}{2}K(\phi_2^2 - \phi_1^2)}{L}x + \frac{1}{2}K\phi_1^2 \quad C_u = 0$$

$$\phi^2 = \frac{\phi_2^2 - \phi_1^2}{L}x + \phi_1^2 \qquad 1.4(12)$$

This equation shows that the phreatic surface varies parabolically with distance (Dupuit's parabola).

The discharge Q_x is now

$$Q_x = -\frac{\partial \Phi}{\partial x} = \frac{\Phi_1 - \Phi_2}{L} \qquad 1.4(13)$$

or

$$Q_x = \frac{K(\phi_1^2 - \phi_2^2)}{2L} \qquad 1.4(14)$$

Radial Flow

In the case of radial flow in an unconfined aquifer as shown in Figure 1.4.3, the results obtained for confined flow can be directly applied to unconfined flow because the governing equations are the same in terms of the discharge potential. From Equation 1.3(31), the governing equation for radial unconfined flow is

$$\Phi = \frac{Q}{2\pi}\ln\left(\frac{r}{R}\right) + \Phi_o \qquad 1.4(15)$$

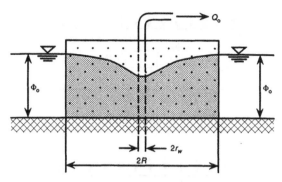

FIG. 1.4.3 Radial flow in an unconfined aquifer.

The governing equation in terms of the head ϕ is

$$\frac{1}{2}K\phi^2 = \frac{Q}{2\pi}\ln\left(\frac{r}{R}\right) + \frac{1}{2}K\phi_o^2$$

$$\phi^2 = \frac{Q}{\pi K}\left(\frac{r}{R}\right) + \phi_o^2 \qquad 1.4(16)$$

or

$$\phi = \sqrt{\frac{Q}{\pi K}\ln\left(\frac{\rho}{P}\right) + \phi_o^2} \qquad 1.4(17)$$

Note that the expression for the head ϕ for radial unconfined flow is different from that for radial confined flow even though the discharge potential for both types of flow is the same. Also, the principle of superposition applies to Φ but not to ϕ. Superposition of two solutions in Equation 1.4(15), therefore, is allowed, but not in Equation 1.4(17).

The introduction of the drawdown s as $s = \phi_o - \phi$ means $\phi^2 = (\phi_o - s)^2 = \phi_o^2 - 2\phi_o s + s^2 = \phi_o^2 - 2\phi_o s (1 - s/2\phi_o)$. Hence, Equation 1.4(16) can be written as

$$s\left(1 - \frac{s}{2\phi_o}\right) = -\frac{Q}{2\pi K\phi_o}\ln\left(\frac{r}{R}\right) \qquad 1.4(18)$$

If drawdown s is small compared to ϕ_o, then $s/2\phi_o \approx 0$, and Equation 1.4(18) can be written as

$$s = \frac{Q}{2\pi K\phi_o}\ln\left(\frac{R}{r}\right) \quad s \ll \phi_o \qquad 1.4(19)$$

This equation is identical to the drawdown equation for confined flow, Equation 1.3(15). This fact is true only if the drawdown is small compared to the head ϕ_o. However, Equation 1.4(19) can be accurate enough as a first approximation.

Unconfined Flow with Infiltration

Water can infiltrate into an unconfined aquifer through the soil above the phreatic surface as the result of rainfall or artificial infiltration. As shown in Figure 1.4.4, water percolates downward into the acquifer at a constant infiltration rate of N per unit area and per unit time.

The continuity equation for unconfined flow, Equation 1.4(5), can be modified to read

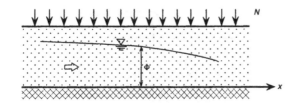

FIG. 1.4.4 Unconfined flow with rainfall.

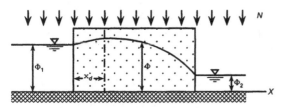

FIG. 1.4.5 One-dimensional unconfined flow with rainfall. (Reprinted from A. Verrjuit, 1982, *Theory of groundwater flow*, Macmillan Pub. Co.)

$$\frac{\partial Q_x}{\partial x} + \frac{\partial Q_y}{\partial y} - N = 0 \qquad 1.4(20)$$

Hence, the differential equation for the potential becomes

$$\frac{\partial^2 \Phi}{\partial x^2} + \frac{\partial^2 \Phi}{\partial y^2} + N = 0 \qquad 1.4(21)$$

In terms of ϕ, this equation reads

$$\frac{\partial^2 \phi}{\partial x^2} + \frac{\partial^2 \phi}{\partial y^2} + \frac{2N}{K} = 0 \qquad 1.4(22)$$

One-Dimensional Flow with Infiltration

For one-dimensional flow shown in Figure 1.4.5, Equation 1.4(21) becomes

$$\frac{d^2 \Phi}{dx^2} + N = 0 \qquad 1.4(23)$$

The general solution of this equation is

$$\Phi = -\frac{N}{2} x^2 + Ax + B \qquad 1.4(24)$$

Use of the boundary condition that

$$x = 0 \qquad \Phi = \Phi_1 = \frac{1}{2} K\phi_1^2$$

$$x = L \qquad \Phi = \Phi_2 = \frac{1}{2} K\phi_2^2$$

gives

$$\Phi = -\frac{N}{2}(x^2 - Lx) - \frac{\Phi_1 - \Phi_2}{L} x + \Phi_1 \qquad 1.4(25)$$

and

$$Q_x = -\frac{d\Phi}{dx} = Nx - \frac{NL}{2} + \frac{\Phi_1 - \Phi_2}{L} \qquad 1.4(26)$$

The location of the divide x_d, where ϕ is maximum, is obtained from

$$\frac{d\Phi}{dx} = 0 = Nx_d - \frac{NL}{2} + \frac{\Phi_1 - \Phi_2}{L} = 0$$

$$\therefore x_d = \frac{\Phi_1 - \Phi_2}{NL} + \frac{L}{2} \qquad (0 \le x_d \le L) \qquad 1.4(27)$$

Note that x_d could be larger than L or could be negative. In those cases, the divide does not exist, and the flow occurs in one direction throughout the aquifer.

Radial Flow with Infiltration

Figure 1.4.6 shows radial flow in an unconfined aquifer with infiltration. If a cylinder has a radius r, the amount of water infiltrating into the cylinder is equal to $Q_{in} = N\pi r^2$, and the amount of water flowing out of the cylinder is equal to $2\pi r \cdot hq_r = 2\pi r Q_r$. The continuity of flow requires that $2\pi r Q_r = N\pi r^2$, giving

$$Q_r = \frac{N}{2} r \qquad 1.4(28)$$

which can be written as

$$Q_r = -\frac{\partial \Phi}{\partial r} = \frac{N}{2} r \qquad 1.4(29)$$

yielding

$$\Phi = -\frac{N}{4} r^2 + C \qquad 1.4(30)$$

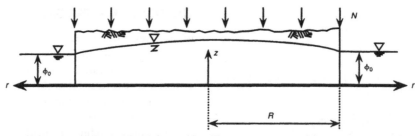

FIG. 1.4.6 Radial unconfined flow with infiltration. (Reprinted from O.D.L. Strack, 1989, *Groundwater mechanics*, Vol. 3, Pt. 3, Prentice-Hall, Inc.)

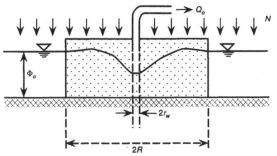

FIG. 1.4.7 Radial flow from pumping with infiltration. (Reprinted from A. Verrjuit, 1982, *Theory of groundwater flow*, 2d ed., Macmillan Pub. Co.)

The constant C in this equation can be determined from the boundary condition that r = R, $\Phi = \Phi_o$. The expression for Φ then becomes

$$\Phi = -\frac{N}{4}(r^2 - R^2) + \Phi_o \qquad 1.4(31)$$

The location of the divide is obviously at the center of the island where $d\Phi/dr = 0$ and $r_d = 0$.

Radial Flow from Pumping Infiltration

Figure 1.4.7 shows radial flow in an unconfined aquifer with infiltration in which water is pumped out of a well located at the center of a circular island.

The principle of superposition can be used to solve this problem. In the first case, the radial flow is from pumping alone; in the second, the flow is from infiltration. Since the differential equations for both cases are linear (Laplace's equation and Poisson's equation), the solution for each can be superimposed to obtain a solution for the whole with the sum of both solutions meeting the boundary conditions.

The addition of the two solutions, Equations 1.4(15) and 1.4(31), with a new constant C gives

$$\Phi = -\frac{N}{4}(r^2 - R^2) + \frac{Q}{2\pi}\ln\left(\frac{r}{R}\right) + C \qquad 1.4(32)$$

The constant C can be obtained from the boundary condition r = R, $\Phi = \Phi_o$. Hence,

$$\Phi = -\frac{N}{4}(r^2 - R^2) + \frac{Q}{2\pi}\ln\left(\frac{r}{R}\right) + \Phi_o \qquad 1.4(33)$$

The discharge Q_r is now obtained as

$$Q_r = -\frac{\partial\Phi}{\partial r} = \frac{N}{2}r - \frac{Q}{2\pi r} \qquad 1.4(34)$$

The divide r_d is a circle and occurs when $Q_r = \partial\Phi/\partial r = 0$ as

$$\frac{N}{2}r_d - \frac{Q}{2\pi r_d} = 0$$

$$\therefore \quad r_d = \sqrt{\frac{Q}{\pi N}} \quad (r_d \le R) \qquad 1.4(35)$$

—*Y.S. Chae*

1.5
COMBINED CONFINED AND UNCONFINED FLOW

As water flows through a confined aquifer, the flow changes from confined to unconfined when the piezometric head ϕ becomes less than the aquifer thickness H. This case is shown in Figure 1.5.1. At the interzonal boundary, the head ϕ becomes equal to the thickness H. The continuity of flow requires no change in discharge at the interzonal boundary. Hence, the following equation governing the discharge potential is the same throughout the flow region:

$$\frac{\partial^2\Phi}{\partial x^2} + \frac{\partial^2\Phi}{\partial y^2} = 0 \qquad 1.5(1)$$

where

$$\Phi = KH\phi + C_c \qquad \text{for} \qquad \phi \ge H$$

$$\Phi = \frac{1}{2}K\phi^2 + C_u \qquad \text{for} \qquad \phi < H$$

At the interzonal boundary, Φ yields the same value, giving

$$KH^2 + C_c = \frac{1}{2}KH^2 + C_u$$

$$C_c = C_u - \frac{1}{2}KH^2 \qquad 1.5(2)$$

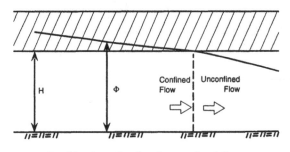

FIG. 1.5.1 Combined confined and unconfined flow.

If one of the two constants C_u is set to zero, then

$$C_c = -\frac{1}{2}KH^2, \qquad C_u = 0 \qquad 1.5(3)$$

The potential Φ can be expressed as

$$\Phi = KH\phi - \frac{1}{2}KH^2 \qquad (\phi \geq H)$$

$$\Phi = \frac{1}{2}K\phi^2 \qquad (\phi < H) \qquad 1.5(4)$$

One-Dimensional Flow

Figure 1.5.2 shows combined confined and unconfined flow in an aquifer of thickness H and length L. The aquifer is confined at $x = 0$ and unconfined at $x = L$.

The expression for the potential Φ is the same throughout the flow region as

$$\Phi = -(\Phi_1 - \Phi_2)\frac{x}{L} + \Phi_1 \qquad 1.5(5)$$

However, the expression for Φ in terms of ϕ is different for each zone as given in Equation 1.5(4). The expression for the discharge Q is

$$Q_x = \frac{\Phi_1 - \Phi_2}{L} = \frac{KH\phi_1 - \frac{1}{2}KH^2 - \frac{1}{2}K\phi_2^2}{L} \qquad 1.5(6)$$

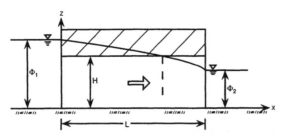

FIG. 1.5.2 One-dimensional combined flow.

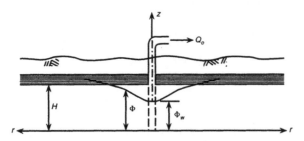

FIG. 1.5.3 Radial combined flow. (Reprinted from O.D.L. Strack, 1989, *Groundwater mechanics*, Vol. 3, Pt. 3, Prentice Hall, Inc.)

The location of the interzonal boundary x_b is obtained from Equation 1.5(5) when Equation 1.5(4) is substituted for Φ_1 and Φ_2, and $\Phi = 1/2\ KH^2$ as

$$x_b = \frac{H\phi_1 - H^2}{H\phi_1 - \frac{1}{2}H^2 - \frac{1}{2}\phi_2^2} \cdot L \qquad 1.5(7)$$

Note that x_b is independent of the hydraulic conductivity K. Also note that when $\phi_1 = H$, $x_b = 0$ (entirely unconfined flow) and when $\phi_2 = H$, $x_b = 1$ (entirely confined flow).

Radial Flow

If the drawdown near the well caused by pumping dips below the aquifer thickness H, then unconfined flow occurs in that region as shown in Figure 1.5.3. The expression for the potential Φ is the same for the entire flow region as

$$\Phi = \frac{Q}{2\pi}\ln\left(\frac{r}{R}\right) + \Phi_o \qquad 1.5(8)$$

In this equation, Φ_o is the potential at $r = R$ when the flow is confined. Hence

$$\Phi_o = KH\phi_o - \frac{1}{2}KH^2 \qquad (\phi_o > H) \qquad 1.5(9)$$

The potential at well Φ_w for unconfined flow is

$$\Phi_w = \frac{1}{2}K\phi_w^2 \qquad (\phi_w < H) \qquad 1.5(10)$$

Equation 1.5(8) can now be rewritten as

$$\frac{1}{2}K\phi_w^2 = \frac{Q}{2\pi}\ln\left(\frac{r_w}{R}\right) + KH\phi_o - \frac{1}{2}KH^2 \qquad 1.5(11)$$

and solving for Q gives

$$Q = \frac{2\pi\left(\frac{1}{2}K\phi_w^2 - KH\phi_o + \frac{1}{2}KH^2\right)}{\ln\left(\frac{r_w}{R}\right)} \qquad 1.5(12)$$

The distance r_b to the interzonal boundary, which is a circle, can be obtained from Equation 1.5(8) with $\Phi = 1/2\ KH^2$ as

$$\frac{1}{2}KH^2 = \frac{Q}{2\pi}\ln\left(\frac{r_b}{R}\right) + \Phi_o$$

$$\therefore \quad r_b = R \cdot e^{2\pi KH(H-\phi_o)/Q} \qquad 1.5(13)$$

—Y.S. Chae

2
Hydraulics of Wells

2.1
TWO-DIMENSIONAL PROBLEMS

This section describes methods for handling two-dimensional groundwater flow problems including superposition, the method of images, and the potential and flow function.

Superposition

The differential equation for two-dimensional steady flow in a homogeneous aquifer is

$$\frac{\partial^2 \Phi}{\partial x^2} + \frac{\partial^2 \Phi}{\partial y^2} = 0 \qquad 2.1(1)$$

Because this equation is a linear and homogeneous differential equation, the principle of superposition applies. The principle states that if two different functions Φ_1 and Φ_2 are solutions of Laplace's equation, then the function

$$\Phi(x,y) = c_1\Phi_1(x,y) + c_2\Phi(x,y) \qquad 2.1(2)$$

is also a solution.

Superposition of solutions is valuable in several groundwater problems. For example, the case of groundwater flow due to simultaneous pumping from several wells can be solved by the superposition of the elementary solution for a single well.

A TWO-WELL SYSTEM

Consider the case of two wells in an infinite aquifer as shown in Figure 2.1.1, in which water is discharged (positive Q) from well 1 and is recharged (negative Q) into well 2. This case is referred to as a sink-and-source problem.

The potential Φ at a point which is located at a distance r_1 from well 1 and r_2 from well 2 can be expressed when the potential Φ_1 is superimposed with respect to well 1 and Φ_2 is superimposed with respect to well 2 as

$$\Phi = \Phi_1 + \Phi_2 = \frac{Q_1}{2\pi} \ell n\, r_1 - \frac{Q_2}{2\pi} \ell n\, r_2 + C \qquad 2.1(3)$$

The constant $C = \Phi_0$ @ $r_1 = r_2 = R$
If $Q_1 = Q_2 = Q$ in a special case, then

$$\Phi = \frac{Q}{2\pi} \ell n\left(\frac{r_1}{r_2}\right) + \Phi_0 \qquad 2.1(4)$$

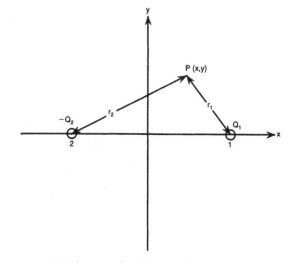

FIG. 2.1.1 Discharge and recharge wells.

or

$$\phi = \frac{Q}{2\pi T} \ell n\left(\frac{r_1}{r_2}\right) + \phi_0 \qquad \text{for a confined aquifer} \qquad 2.1(5)$$

$$\phi^2 = \frac{Q}{\pi K} \ell n\left(\frac{r_1}{r_2}\right) + \phi_0^2 \qquad \text{for an unconfined aquifer} \qquad 2.1(6)$$

Figure 2.1.2 shows the flow net for a two-well sink-and-source system. Equation 2.1(4) shows that along the y axis where $r_1 = r_2 = r_0$, Φ = constant. This statement means that the y axis is an equipotential line along which no flow occurs, and the drawdown is zero ($\phi = \phi_0$). This result occurs because the system is in symmetry about the y axis and the problem is linear. Note that the distance R does not appear in Equation 2.1(4). This omission is because the discharge from the sink is equal to the recharge into the source, indicating that the system is in hydraulic equilibrium requiring no external supply of water.

Another example of using the principle of superposition is the case of two sinks of equal discharge Q. Equation 2.1(3) now reads

$$\Phi = \Phi_1 + \Phi_2 = \frac{Q}{2\pi} \ell n\,(r_1 r_2) + C \qquad 2.1(7)$$

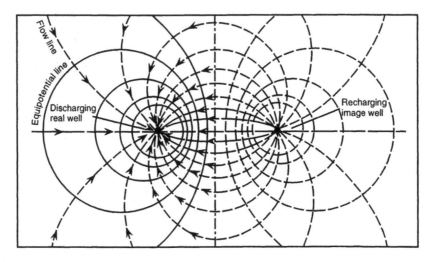

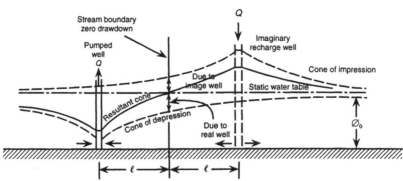

FIG. 2.1.2 Source and sink in unconfined flow. (Reprinted from R.S. Gupta, 1989, *Hydrology and hydraulic systems,* Prentice-Hall, Inc.)

Use of the boundary condition $r = R$, $\Phi = \Phi_o$ yields

$$\Phi = \frac{Q}{2\pi} \ell n\left(\frac{r_1 r_2}{R^2}\right) + \Phi_0 \qquad \text{2.1(8)}$$

Figure 2.1.3 shows the flow net for a two-well sink-and-sink system. The y axis plays the role of an impervious boundary along which no water flows across. This result occurs because the flow at points on the y axis is directed along the axis due to the equal pull of flow from the two wells located equidistance from the points.

A MULTIPLE-WELL SYSTEM

The principle of superposition previously discussed for two wells can be applied to a system of multiple wells, n wells in number from $i = 1$ to n. The solution for such a system can be written with the use of superposition as

$$\Phi = \frac{1}{2\pi} \left[\sum_{i=i}^{n} Q_i \ell n\left(\frac{r_i}{R}\right)\right] + \Phi_o \qquad \text{2.1(9)}$$

or

$$\phi = \frac{1}{2\pi T} \left[\sum_{i=1}^{n} Q_i \ell n\left(\frac{r_i}{R}\right)\right] + \phi_o \quad \text{for a confined aquifer}$$

$$\text{2.1(10)}$$

and

$$\phi^2 = \frac{1}{\pi K} \left[\sum_{i=1}^{n} Q_i \ell n\left(\frac{r_i}{R}\right)\right] + \phi_o^2 \quad \text{for an unconfined aquifer}$$

$$\text{2.1(11)}$$

The drawdown at the jth well is then

$$\phi_w = \frac{1}{2\pi T} \left[Q_j \ell n\left(\frac{r_w}{R}\right) + \sum_{i=1}^{n-1} Q_i \ell n\left(\frac{r_{i,j}}{R}\right)\right] + \phi_o$$

$$\text{for a confined aquifer} \quad \text{2.1(12)}$$

and

$$\phi_w^2 = \frac{1}{\pi K} \left[Q_j \ell n\left(\frac{r_w}{R}\right) + \sum_{i=1}^{n-1} Q_i \ell n\left(\frac{r_{i,j}}{R}\right)\right] + \phi_o^2$$

$$\text{for an unconfined aquifer} \quad \text{2.1(13)}$$

where $r_{i,j}$ is the distance between the jth well and ith wells. The quantities inside the brackets [ ] in these equations are called the *drawdown factors,* F_p at a point and F_w at a well, respectively. These equations can be rewritten as

$$\Phi = \Phi_o + \frac{1}{2\pi} F_p \quad \text{at a point} \qquad \text{2.1(14)}$$

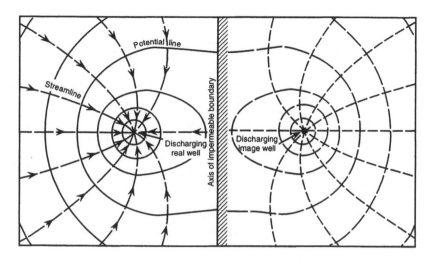

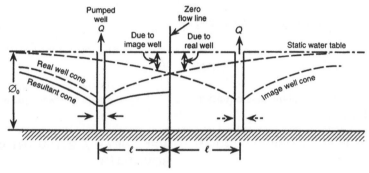

FIG. 2.1.3 Sink and sink in unconfined flow. (Reprinted from R.S. Gupta, 1989, *Hydrology and hydraulic systems,* Prentice-Hall, Inc.)

$$\Phi_w = \Phi_o + \frac{1}{2\pi} F_w \qquad \text{at a well} \qquad 2.1(15)$$

where

$$F_p = \sum_{i=1}^{n} Q_i \, \ell n \left(\frac{r_w}{R}\right) \qquad 2.1(16)$$

$$F_w = Q_j \, \ell n \left(\frac{r_w}{R}\right) + \sum_{i=1}^{n-1} Q_i \, \ell n \left(\frac{r_{i,j}}{R}\right) \qquad 2.1(17)$$

The following examples give the drawdown factors of wells in special arrays:

a. Circular array, n wells in equal spacing (Figure 2.1.4a)

$$F_p = nQ \, \ell n \frac{\rho}{R} \qquad 2.1(18)$$

$$F_w = Q \, \ell n \frac{nr_w \rho^{n-1}}{R^n} \qquad 2.1(19)$$

b. Rectangular array (Figure 2.1.4b)
   • Approximate method:
       Equivalent radius $\rho_e = 4\sqrt{ab}/\pi$
       Then use Equation 2.1(11)
   • Exact method:
       Use Equation 2.1(9), 2.1(12) or 2.1(13)

c. Two parallel lines of equally spaced wells (Figure 2.1.4c)

$$F_c = 4Q \sum_{i=1}^{i=n/4} \ell n \frac{R}{\frac{1}{2} \cdot \sqrt{S^2(2i-1)^2 + B^2}} \qquad 2.1(20)$$

$$F_w = 2Q \sum_{i=1}^{i=n/2} \ell n \frac{R}{\frac{1}{2} \cdot \sqrt{S^2(2i-3)^2 + B^2}} \qquad 2.1(21)$$

Method of Images

A special application of superposition is the method of images. This method can be used to solve problems involving the flow in aquifers of relatively simple geometrical form such as an infinite strip, a half plane, or a quarter plane. The following problems are specific examples.

WELL NEAR A STRAIGHT RIVER

To solve the problem of a well near a long body of water (river, canal, or lake) shown in Figure 2.1.5, replace the half-plane aquifer by an imaginary infinite aquifer with an imaginary well placed at the mirror image position from the real well. This case now represents the sink and source problem discussed previously, and Equation 2.1(4) satis-

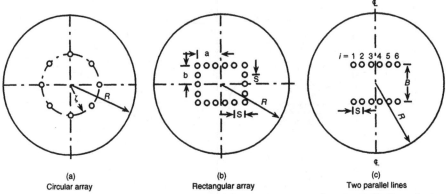

(a)
Circular array

(b)
Rectangular array

(c)
Two parallel lines

FIG. 2.1.4 Wells in special arrays. (Reprinted from G.A. Leonards, ed., 1962, *Foundation engineering*, McGraw-Hill, Inc.)

fies all conditions associated with the case. Accordingly, the solution is given by

$$\Phi = \frac{Q}{2\pi} \ln\left(\frac{r_1}{r_2}\right) + \Phi_o \qquad 2.1(22)$$

If n number of wells are on the half plane, use Equation 2.1(7) for solution as follows:

$$\Phi = \Phi_o + \frac{1}{2\pi} F'_p \quad \text{at a point} \qquad 2.1(23)$$

$$\Phi_w = \Phi_o + \frac{1}{2\pi} F'_w \quad \text{at a well} \qquad 2.1(24)$$

where

$$F'_p = \sum_{i=1}^{n} Q_i \, \ell n\left(\frac{r_i}{r'_i}\right) \qquad 2.1(25)$$

$$F'_w = Q_i \, \ell n\left(\frac{r_w}{r'_j}\right) + \sum_{i=1}^{n-1} Q_i \, \ell n\left(\frac{r_{i,j}}{r'_{i,j}}\right) \qquad 2.1(26)$$

and

r'_i = distance between point and imaginary ith well.
$r'_{i,j}$ = distance between jth well and imaginary ith well.

WELL NEAR A STRAIGHT IMPERVIOUS BOUNDARY

The problem of a well near a long straight impervious boundary (e.g. a mountain ridge or fault) is solved in a similar manner as that of a well near a straight river. In this case, the type of image well is a sink rather than a source as shown in Figure 2.1.6.

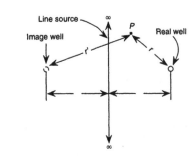

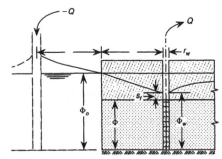

FIG. 2.1.5 Well near a straight river. (Reprinted from G.A. Leonards, ed., 1962, *Foundation engineering*, McGraw-Hill, Inc.)

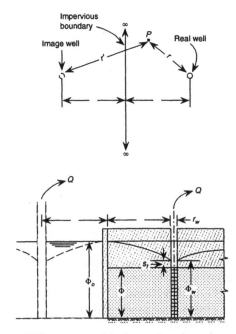

FIG. 2.1.6 Well near a straight impervious boundary.

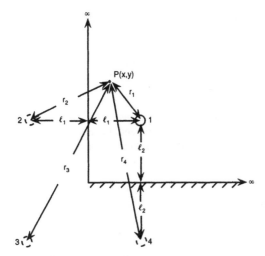

FIG. 2.1.7 Well in a quarter plane.

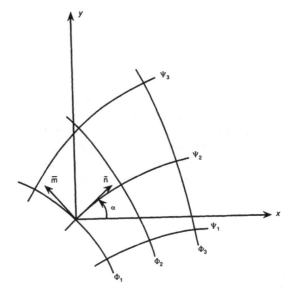

FIG. 2.1.8 Potential and flow lines.

The solution for this case, therefore, is the same as the case of a sink-and-sink problem given by Equation 2.1(8), which is

$$\Phi = \frac{Q}{2\pi} \ell n \left(\frac{r_1 r_2}{R}\right) + \Phi_o \qquad 2.1(27)$$

WELL IN A QUARTER PLANE

Figure 2.1.7 shows the case of a well operating in an aquifer bounded by a straight river and an impervious boundary. To solve this problem, place a series of imaginary wells (wells numbered 2, 3, and 4), and use superposition. Figure 2.1.7 indicates that wells 2 and 3 are sources, and well 4 is a sink. Hence,

$$\Phi = \frac{Q}{2\pi}\left[\ell n\left(\frac{r_1}{R}\right) - \ell n\left(\frac{r_2}{R}\right) - \ell n\left(\frac{r_3}{R}\right) + \ell n\left(\frac{r_4}{R}\right) \right] + \Phi_o$$

$$2.1(28)$$

Potential and Flow Functions

In the Basic Equations section, the fundamental equation of groundwater flow expressed in terms of discharge potential Φ is:

$$\frac{\partial^2 \Phi}{\partial x_2} + \frac{\partial^2 \Phi}{\partial y^2} = 0 \qquad 2.1(29)$$

The potential $\Phi(x,y)$ is a single-value function everywhere in the x, y plane. Therefore, lines of constant Φ_1, Φ_2, ..., called *equipotential lines,* can be drawn in the x, y plane as shown in Figure 2.1.8. When the lines are drawn with a constant interval between the values of the two successive lines ($\Delta\Phi = \Phi_1 - \Phi_2 = \Phi_2 - \Phi_3 = ...$), then an equal and constant amount of potential drop is between any two of the equipotential lines.

At any arbitrary point on the equipotential line, flow occurs only in the direction perpendicular to the line (n direction), and no flow occurs in the tangential direction (m direction) as

$$q_n = \frac{\partial \Phi}{\partial \bar{n}} = -q; \qquad q_m = \frac{\partial \Phi}{\partial \bar{m}} = 0 \qquad 2.1(30)$$

Accordingly, lines can be drawn perpendicular to the equipotential lines as shown in Figure 2.1.8. These lines are called flow or stream lines.

At this point a second function, called flow or stream function Ψ, is introduced. Since the specific discharge vector must satisfy the equation of continuity, the function Ψ is defined by

$$q_x = -\frac{\partial \Psi}{\partial y}, \qquad q_y = \frac{\partial \psi}{\partial x} \qquad 2.1(31)$$

It now follows that

$$\frac{\partial^2 \Psi}{\partial x^2} + \frac{\partial^2 \Psi}{\partial y^2} = 0 \qquad 2.1(32)$$

or

$$\nabla^2 \psi = 0 \qquad 2.1(33)$$

which shows that Ψ is, like the potential Φ, a harmonic function and should satisfy Laplace's equation.

The directional function Ψ with respect to m and n can now be easily written as follows because no flow component is in m direction:

$$\frac{\partial \psi}{\partial \bar{m}} = -q, \qquad \frac{\partial \psi}{\partial \bar{n}} = 0 \qquad 2.1(34)$$

The lines of constant Φ and Ψ form a set of orthogonal curves called a flow net. Also,

$$q = -\frac{\partial \Phi}{\partial \bar{n}} = -\frac{\partial \psi}{\partial \bar{m}} \qquad 2.1(35)$$

meaning that if lines are drawn with constant Φ and Ψ at intervals $\Delta\Phi$ and $\Delta\psi$, then

$$\frac{\Delta\Phi}{\Delta\bar{n}} = \frac{\Delta\psi}{\Delta\bar{m}} \qquad 2.1(36)$$

where Δn is the distance between two potential lines, and Δm is the distance between two flow lines. Thus, the equipotential lines and flow lines are not only orthogonal, but they form elementary curvilinear squares. This property is the basis of using a flow net as an approximate graphic method to solve groundwater problems. With a flow net drawn, for example, the rate of flow (Q) can be obtained by

$$Q = K\phi_o \frac{n_f}{n_\phi} \qquad 2.1(37)$$

where:

    n_f = number of flow zones
    n_ϕ = number of equipotential zones
    ϕ_o = total head loss in flow system

—Y.S. Chae

2.2
NONSTEADY (TRANSIENT) FLOW

Nonsteady or transient flow in aquifers occurs when the pressure and head in the aquifer change gradually until steady-state conditions are reached. During the course of transient flow, water can be either stored in or released from the soil. Storage has two possibilities. First, water can simply fill the pore space in soil without changing the soil volume. This storage is called *phreatic storage,* and usually occurs in unconfined aquifers as the groundwater table moves up or down. In the other storage, water is stored in the pore space increased by deformation of the soil and involves a volume change. This storage is called *elastic storage* and occurs in all types of aquifers. However, in confined aquifers, it is the only form of storage.

Transient Confined Flow (Elastic Storage)

In a completely saturated confined aquifer, water can be stored or released if the change in aquifer pressure results in volumetric deformation of the soil. The problem is complex because the constitutive equations for soil are highly nonlinear even for dry soil, and coupling them with groundwater flow increases the complexity.

The basic equation for the phenomenon is the storage equation (Strack 1989), as

$$\nabla^2\Phi = H\left[\frac{\partial\varepsilon_o}{\partial t} + n\beta\frac{\partial p}{\partial t}\right] \qquad 2.2(1)$$

where ε_o = volume strain, and β = compressibility of water. From soil mechanics

$$\frac{\partial\varepsilon_o}{\partial t} = m_v\frac{\partial p}{\partial t} \qquad 2.2(2)$$

where m_v = modulus of volume change. Equation 2.2(1) can then be written as

$$\nabla^2\Phi = H(m_v + n\beta)\frac{\partial p}{\partial t} \qquad 2.2(3)$$

When the variation of K, H, and ρ with time are neglected, then

$$\frac{\partial\Phi}{\partial t} = KH\frac{\partial}{\partial t}\left[\frac{p}{\rho g} + z\right] = \frac{KH}{\rho g}\frac{\partial p}{\partial t} \qquad 2.2(4)$$

so that Equation 2.2(3) can be written as

$$\nabla^2\Phi = \frac{S_s}{K}\frac{\partial\Phi}{\partial t} \qquad 2.2(5)$$

or

$$\frac{\partial\Phi}{\partial t} = \frac{K}{S_s}\nabla^2\Phi \qquad 2.2(6)$$

where S_s [(1/m)] is the coefficient of specific storage

$$S_s = \rho g(m_v + n\beta) \qquad 2.2(7)$$

If the compressibility of water β is ignored, then $S_s = \rho g m_v$.

Some typical values of m_v are given in Table 2.2.1. Equation 2.2(5) can also be written in terms of ϕ as

$$\nabla^2\phi = \frac{S_e}{T}\frac{\partial\phi}{\partial t} \qquad 2.2(8)$$

where S_e = coefficient of elastic storage = $S_s \cdot H$.

Transient Unconfined Flow (Phreatic Storage)

The vertical movement of a phreatic surface results in water being stored in soil pores without causing the soil to

TABLE 2.2.1 TYPICAL VALUES OF COMPRESSIBILITY (m_v)

	Compressibility, (m^2/N or Pa^{-1})
Clay	10^{-6}–10^{-8}
Sand	10^{-7}–10^{-9}
Gravel	10^{-8}–10^{-10}
Jointed rock	10^{-8}–10^{-10}
Sound rock	10^{-9}–10^{-11}
Water (β)	4.4×10^{-10}

Source: R.A. Freeze and J.A. Cherry, 1979, *Groundwater* (Prentice-Hall, Inc.).

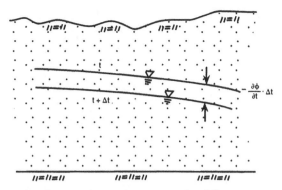

FIG. 2.2.1 Storage change due to unconfined flow.

deform. Phreatic storage is, therefore, several orders of magnitude greater than elastic storage, which can be ignored.

The basic differential equation for the transient unconfined flow (Strack 1989), such as shown in Figure 2.2.1, can be given as

$$\nabla^2 \Phi = S_p \frac{\partial \phi}{\partial t} \qquad 2.2(9)$$

where S_p = coefficient of phreatic storage.

Equation 2.2(9) can be linearized in terms of the potential Φ as

$$\nabla^2 \Phi = \frac{S_s}{K} \frac{\partial \Phi}{\partial t} \qquad 2.2(10)$$

or

$$\frac{\partial \Phi}{\partial t} = \frac{K}{S_s} \nabla^2 \Phi \qquad 2.2(11)$$

This equation is the same as that for transient confined flow. However, S_s is related to S_p as $S_s = S_p/\overline{\phi}$, where $\overline{\phi}$ is the average piezometric head in the aquifer.

Transient Radial Flow (Theis Solution)

The governing equation for the transient radial flow (flow toward a well in an aquifer of infinite extent) is obtained when Equation 2.2(10) is rewritten in terms of radial coordinate r as

$$\frac{\partial^2 \Phi}{\partial r^2} + \frac{1}{r} \frac{\partial \Phi}{\partial r} = \frac{S_s}{K} \frac{\partial \Phi}{\partial t} \qquad 2.2(12)$$

The solution to this equation is commonly given as

$$\Phi = -\frac{Q}{4\pi} \text{Ei}(u) + \Phi_o \qquad 2.2(13)$$

known as the Theis solution. Ei is the exponential integral, and u is a dimensionless variable defined by

$$u = \frac{S_s r^2}{4Kt} \qquad 2.2(14)$$

or

$$u = \frac{S_e r^2}{4Tt} \qquad \text{for confined flow} \qquad (T = KH) \qquad 2.2(15)$$

or

$$u = \frac{S_p r^2}{4Tt} \qquad \text{for unconfined flow} \qquad (T = K\overline{\phi}) \qquad 2.2(16)$$

The exponential integral Ei(u) is referred to as the well function W(u). Ei(u) can be approximated by

$$\text{Ei}(u) = -\left[0.577216 - \ell n\, u + u - \frac{u^2}{2.2!} + \frac{u^3}{3.3!} - \cdots \right] \qquad 2.2(17)$$

Using the well function W(u), the Theis solution can be written as

$$\Phi = -\frac{Q}{4\pi} W(u) + \Phi_o \qquad 2.2(18)$$

or in terms of the head ϕ as

$$\phi = -\frac{Q}{4\pi T} W(u) + \phi_o \qquad 2.2(19)$$

The drawdown s is obtained by

$$s = \frac{Q}{4\pi T} W(u) \qquad 2.2(20)$$

Values of W(u) for different values of u are shown in Table 2.2.2. The drawdown s at a given distance r from the well at given time t can be calculated from Equation 2.2(20) and Table 2.2.2.

Figure 2.2.2, accompanied by Table 2.2.3, shows an example of drawdown versus a time curve for a transient radial flow in a confined aquifer with T = 1000 m^2/d and S = 0.0001 for a pumping rate of Q = 1000 m^3/d. The figure shows that even in a transient flow, the rate of drawdown (Δs) achieves a steady state after a short period of pumping, two days in this example.

If u is small (e.g., less than 0.01), only the first two terms of the brackets in Equation 2.2(17) are significant. Equation 2.2(19) can be simplified to

TABLE 2.2.2 VALUES OF W(U) FOR DIFFERENT VALUES OF U

u \ N	N	$N\times10^{-1}$	$N\times10^{-2}$	$N\times10^{-3}$	$N\times10^{-4}$	$N\times10^{-5}$	$N\times10^{-6}$	$N\times10^{-7}$	$N\times10^{-8}$	$N\times10^{-9}$	$N\times10^{-10}$	$N\times10^{-11}$	$N\times10^{-12}$	$N\times10^{-13}$	$N\times10^{-14}$	$N\times10^{-15}$
1	0.219	1.82	4.04	6.33	8.63	10.9	13.2	15.5	17.8	20.1	22.4	24.8	27.1	29.4	31.7	34.0
1.2	0.158	1.66	3.86	6.15	8.45	10.8	13.1	15.4	17.7	20.0	22.3	24.6	26.9	29.2	31.5	33.8
1.5	0.100	1.46	3.64	5.93	8.23	10.5	12.8	15.1	17.4	19.7	22.0	24.3	26.6	29.0	31.3	33.6
2	0.0489	1.22	3.35	5.64	7.94	10.2	12.5	14.8	17.2	19.5	21.8	24.1	26.4	28.7	31.0	33.3
2.2	0.0372	1.15	3.26	5.54	7.84	10.1	12.4	14.8	17.1	19.4	21.7	24.0	26.3	28.6	30.9	33.2
2.5	0.0249	1.04	3.14	5.42	7.72	10.0	12.3	14.6	16.9	19.2	21.5	23.8	26.1	28.4	30.7	33.0
3	0.0130	0.906	2.96	5.23	7.53	9.84	12.1	14.4	16.7	19.0	21.3	23.7	26.0	28.3	30.6	32.9
3.2	0.0101	0.858	2.90	5.17	7.47	9.77	12.1	14.4	16.7	19.0	21.3	23.6	25.9	28.2	30.5	32.8
3.5	0.00697	0.794	2.81	5.08	7.38	9.68	12.0	14.3	16.6	18.9	21.2	23.5	25.8	28.1	30.4	32.7
4	0.00378	0.702	2.68	4.95	7.25	9.55	11.9	14.2	16.5	18.8	21.1	23.4	25.7	28.0	30.3	32.6
4.2	0.00300	0.670	2.63	4.90	7.20	9.50	11.8	14.1	16.4	18.7	21.0	23.3	25.6	27.9	30.2	32.5
4.5	0.00207	0.625	2.57	4.83	7.13	9.43	11.7	14.0	16.3	18.6	20.9	23.2	25.5	27.9	30.2	32.5
5	0.00115	0.560	2.47	4.73	7.02	9.33	11.6	13.9	16.2	18.5	20.8	23.1	25.4	27.7	30.0	32.4
5.2	0.000909	0.536	2.43	4.69	6.98	9.29	11.6	13.9	16.2	18.5	20.8	23.1	25.4	27.7	30.0	32.3
5.5	0.000641	0.503	2.38	4.63	6.93	9.23	11.5	13.8	16.1	18.4	20.7	23.0	25.3	27.7	30.0	32.3
6	0.000360	0.454	2.30	4.54	6.84	9.14	11.4	13.7	16.1	18.4	20.7	23.0	25.3	27.6	29.9	32.2
6.2	0.000286	0.437	2.26	4.51	6.81	9.11	11.4	13.7	16.0	18.3	20.6	22.9	25.2	27.5	29.8	32.1
6.5	0.000203	0.411	2.22	4.47	6.76	9.06	11.4	13.7	16.0	18.3	20.6	22.9	25.2	27.5	29.8	32.1
7	0.000115	0.374	2.15	4.39	6.69	8.99	11.3	13.6	15.9	18.2	20.5	22.8	25.1	27.4	29.7	32.0
7.2	0.0000922	0.360	2.12	4.36	6.66	8.96	11.3	13.6	15.9	18.2	20.5	22.8	25.1	27.4	29.7	32.0
7.5	0.0000658	0.340	2.09	4.32	6.62	8.92	11.2	13.5	15.8	18.1	20.4	22.7	25.0	27.3	29.6	32.0
8	0.0000377	0.311	2.03	4.26	6.55	8.86	11.2	13.5	15.8	18.1	20.4	22.7	25.0	27.3	29.6	31.9
8.2	0.0000301	0.300	2.00	4.23	6.53	8.83	11.1	13.4	15.7	18.0	20.3	22.6	24.9	27.3	29.6	31.9
8.5	0.0000216	0.284	1.97	4.20	6.49	8.80	11.1	13.4	15.7	18.0	20.3	22.6	24.9	27.2	29.5	31.8
9	0.0000124	0.260	1.92	4.14	6.44	8.74	11.0	13.3	15.6	17.9	20.3	22.6	24.9	27.2	29.5	31.8
9.2	0.00000999	0.251	1.90	4.12	6.41	8.72	11.0	13.3	15.6	17.9	20.2	22.5	24.8	27.1	29.4	31.7
9.5	0.00000718	0.239	1.87	4.09	6.38	8.68	11.0	13.3	15.6	17.9	20.2	22.5	24.8	27.1	29.4	31.7
10		0.219	1.82	4.04	6.33	8.63	10.9	13.2	15.5	17.8	20.1	22.4	24.8	27.1	29.4	31.7

Source: H. Bouwer, 1978, *Groundwater hydrology* (McGraw-Hill, Inc.).

TABLE 2.2.3   CALCULATION OF S IN RELATION TO T

t, Days	r = 100 m			r = 200 m		
	u	W(u)	s, m	u	W(u)	s, m
0.001	0.25	1.044	0.083	1	0.219	0.017
0.005	0.05	2.468	0.196	0.2	1.223	0.097
0.01	0.025	3.136	0.249	0.1	1.823	0.145
0.05	0.005	4.726	0.376	0.02	3.355	0.267
0.1	0.002 5	5.417	0.431	0.01	4.038	0.322
0.5	0.000 5	7.024	0.559	0.002	5.639	0.449
1	0.000 25	7.717	0.614	0.001	6.331	0.504
5	0.000 05	9.326	0.742	0.000 2	7.940	0.632
10	0.000 025	10.019	0.797	0.000 1	8.633	0.687

Source: H. Bouwer, 1978, *Groundwater hydrology* (McGraw-Hill, Inc.).

$$\phi = -\frac{Q}{4\pi T} \ell n \left(\frac{2.25Tt}{Sr^2} \right) + \phi_o \qquad 2.2(21)$$

then

$$s = \frac{Q}{4\pi T} \ell n \left(\frac{2.25Tt}{Sr^2} \right) \qquad 2.2(22)$$

Equation 2.2(21) can be rewritten as

$$\phi = \frac{Q}{2\pi T} \ell n \left[\frac{r}{\left(2.25 \frac{Tt}{S} \right)^{1/2}} \right] + \phi_o \qquad 2.2(23)$$

$$\phi = \frac{Q}{2\pi T} \ell n \left(\frac{r}{R_{eq.}} \right) + \phi_o \qquad 2.2(24)$$

where

$$R_{eq.} = \left(2.25 \frac{Tt}{S} \right)^{1/2}$$

Note that Equation 2.2(24) is similar in expression to the steady-state flow. Equation 2.2(22) allows direct calculation of drawdown in terms of distance r and time t for given aquifer characteristics T and S at a known pumping rate Q.

The exact solution of Equation 2.2(13) is difficult for unconfined aquifers because $\overline{T} = K\overline{\phi}$ is not constant but varies with distance r and time t. The average head $\overline{\phi}$ can be estimated and used in the Theis solution for small drawdowns. For large drawdowns, however, the use of $\overline{\phi}$ for the Theis solution is not valid.

For large drawdowns, Boulton (1954) presents a solution which is valid if the water depth in the well exceeds 0.5 ϕ_o. Boulton's equation is:

$$s = \frac{Q}{2\pi K\phi_o} (1 + C_k)V(t', r') \qquad 2.2(25)$$

where V is Boulton's well function, and C_k is a correction factor. The t' and r' are defined as

$$r' = \frac{1}{\phi_o} \cdot r \qquad 2.2(26)$$

$$t' = \frac{K}{S_p \phi_o} t \qquad 2.2(27)$$

The values of V(r',t') and C_k are given in Table 2.2.4 and Table 2.2.5 respectively.

The head at the well ϕ_w can be calculated from the equation (Bouwer 1988) as

$$Q_w^2 = \phi_o^2 - \frac{Q}{\pi k} \ell n \left(1.5 \sqrt{\frac{Kt}{S_p r_w}} \right) \qquad 2.2(28)$$

which is valid if t' = $(Kt/\phi_o S) > 5$. If t' is smaller than 5, ϕ_w is calculated as

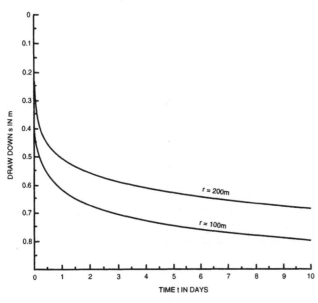

FIG. 2.2.2 Drawdown versus time due to pumping from a well. (Reprinted from H. Bouwer, 1978, *Groundwater hydrology*, McGraw-Hill, Inc.)

TABLE 2.2.4 VALUES OF THE FUNCTION V(T',R') FOR DIFFERENT VALUES OF T' AND R'

t'	0.001	0.002	0.003	0.004	0.005	0.006	0.007	0.008	0.009	0.01	0.02	0.03	0.04	0.05	0.06	0.07	0.08	0.09
0.01	2.99	2.30	1.90	1.64	1.42	1.28	1.15	1.04	0.95	0.875	0.474	0.322	0.240	0.192	0.158	0.135	0.118	0.104
0.02	3.68	2.97	2.58	2.30	2.09	1.92	1.76	1.64	1.52	1.42	0.860	0.610	0.468	0.378	0.316	0.270	0.236	0.210
0.03	4.08	3.40	3.00	2.70	2.46	2.28	2.13	2.00	1.88	1.79	1.18	0.860	0.675	0.555	0.465	0.400	0.350	0.310
0.04	4.35	3.68	3.26	2.98	2.75	2.58	2.42	2.29	2.17	2.06	1.42	1.07	0.850	0.710	0.600	0.525	0.460	0.410
0.05	4.58	3.90	3.49	3.20	2.96	2.79	2.64	2.50	2.38	2.28	1.60	1.24	1.010	0.850	0.725	0.630	0.560	0.500
0.06	4.76	4.06	3.65	3.36	3.15	2.96	2.80	2.68	2.56	2.45	1.78	1.40	1.15	0.970	0.840	0.735	0.650	0.585
0.07	4.92	4.20	3.80	3.51	3.30	3.12	2.96	2.82	2.70	2.60	1.91	1.54	1.28	1.09	0.950	0.835	0.740	0.670
0.08	5.08	4.34	3.94	3.65	3.42	3.24	3.09	2.95	2.84	2.72	2.04	1.65	1.39	1.20	1.04	0.925	0.825	0.750
0.09	5.18	4.47	4.05	3.75	3.54	3.35	3.20	3.05	2.95	2.84	2.14	1.75	1.50	1.29	1.14	1.02	0.910	0.825
0.1	5.24	4.54	4.14	3.85	3.63	3.45	3.30	3.15	3.04	2.94	2.25	1.85	1.58	1.38	1.22	1.09	0.985	0.890
0.2	5.85	5.15	4.78	4.50	4.28	4.10	3.93	3.80	3.66	3.56	2.87	2.46	2.20	1.98	1.80	1.65	1.52	1.42
0.3	6.24	5.50	5.12	4.85	4.61	4.43	4.28	4.14	4.01	3.90	3.24	2.84	2.54	2.32	2.14	1.98	1.85	1.74
0.4	6.45	5.75	5.35	5.08	4.85	4.67	4.50	4.38	4.26	4.15	3.46	3.05	2.76	2.54	2.36	2.20	2.07	1.96
0.5	6.65	6.00	5.58	5.25	5.00	4.85	4.70	4.55	4.45	4.30	3.65	3.24	2.95	2.72	2.52	2.38	2.24	2.14
0.6	6.75	6.10	5.65	5.40	5.15	4.98	4.82	4.68	4.56	4.45	3.76	3.37	3.09	2.85	2.67	2.50	2.38	2.26
0.7	6.88	6.20	5.80	5.50	5.25	5.08	4.92	4.80	4.68	4.55	3.90	3.50	3.20	2.99	2.80	2.64	2.50	2.38
0.8	7.00	6.25	5.85	5.60	5.35	5.20	5.00	4.90	4.80	4.65	3.96	3.55	3.26	3.05	2.86	2.71	2.58	2.46
0.9	7.10	6.35	6.00	5.70	5.50	5.30	5.12	5.00	4.90	4.75	4.05	3.65	3.36	3.15	2.96	2.80	2.66	2.55
1	7.14	6.45	6.05	5.75	5.55	5.35	5.20	5.05	4.95	4.83	4.10	3.74	3.45	3.22	3.04	2.90	2.75	2.64
2	7.60	6.88	6.45	6.15	5.92	5.75	5.60	5.50	5.35	5.25	4.59	4.18	3.90	3.68	3.50	3.34	3.20	3.09
3	7.85	7.15	6.70	6.45	6.20	6.00	5.85	5.75	5.60	5.50	4.82	4.42	4.12	3.90	3.72	3.57	3.45	3.31
4	8.00	7.28	6.85	6.58	6.35	6.15	6.00	5.90	5.75	5.70	4.95	4.55	4.26	4.04	3.86	3.70	3.59	3.46
5	8.15	7.35	7.00	6.65	6.50	6.25	6.10	6.00	5.85	5.80	5.05	4.68	4.40	4.19	4.00	3.85	3.71	3.60
6	8.20	7.50	7.10	6.75	6.55	6.35	6.20	6.10	5.95	5.85	5.20	4.78	4.50	4.26	4.09	3.92	3.80	3.69
7	8.25	7.55	7.15	6.85	6.62	6.40	6.30	6.20	6.05	5.95	5.25	4.85	4.58	4.35	4.18	4.00	3.90	3.78
8	8.30	7.60	7.20	6.90	6.70	6.50	6.35	6.25	6.10	6.05	5.30	4.92	4.65	4.40	4.25	4.10	3.95	3.82
9	8.32	7.65	7.25	7.00	6.75	6.55	6.40	6.30	6.15	6.10	5.35	5.00	4.70	4.49	4.30	4.15	4.00	3.90
10	8.35	7.75	7.35	7.05	6.80	6.60	6.45	6.35	6.20	6.14	5.40	5.02	4.80	4.52	4.35	4.19	4.05	3.92

Continued on next page

TABLE 2.2.4 Continued

t' \ r'	0.1	0.2	0.3	0.4	0.5	0.6	0.7	0.8	0.9	1	2	3	4	5
0.01	0.093	0.0430	0.0264	0.0180	0.0132	0.0100	0.0078	0.0062	0.0049	0.0040	0.00057	0.00015		
0.02	0.187	0.0865	0.0530	0.0365	0.0268	0.0205	0.0160	0.0125	0.0100	0.0081	0.00118	0.00020		
0.03	0.278	0.130	0.0800	0.0550	0.0405	0.0310	0.0240	0.0190	0.0150	0.0122	0.00184	0.00032		
0.04	0.368	0.174	0.107	0.0735	0.0540	0.0415	0.0322	0.0255	0.0202	0.0165	0.00244	0.00043		
0.05	0.450	0.215	0.133	0.0920	0.0675	0.0520	0.0400	0.0320	0.0255	0.0206	0.00305	0.00055		
0.06	0.530	0.257	0.160	0.110	0.0810	0.0610	0.0478	0.0380	0.0305	0.0250	0.00365	0.00065		
0.07	0.610	0.298	0.186	0.130	0.0950	0.0725	0.0565	0.0450	0.0360	0.0292	0.00430	0.00078		
0.08	0.680	0.340	0.214	0.148	0.108	0.0825	0.0645	0.0510	0.0412	0.0336	0.00500	0.00090		
0.09	0.750	0.378	0.236	0.164	0.122	0.0930	0.0730	0.0585	0.0470	0.0380	0.00570	0.00105		
0.1	0.815	0.415	0.260	0.180	0.134	0.103	0.0805	0.0640	0.0515	0.0420	0.00635	0.00118		
0.2	1.32	0.750	0.500	0.359	0.268	0.208	0.165	0.132	0.107	0.0880	0.0145	0.00278		
0.3	1.64	1.02	0.700	0.515	0.392	0.308	0.246	0.200	0.164	0.135	0.0238	0.00490		
0.4	1.86	1.22	0.870	0.650	0.510	0.405	0.328	0.268	0.220	0.182	0.0350	0.00750	0.00160	0.00038
0.5	2.03	1.37	1.00	0.770	0.610	0.490	0.400	0.330	0.275	0.230	0.0450	0.0104	0.00240	0.00056
0.6	2.16	1.49	1.12	0.875	0.700	0.570	0.468	0.390	0.325	0.276	0.0580	0.0138	0.00320	0.00080
0.7	2.28	1.60	1.22	0.965	0.775	0.640	0.525	0.445	0.375	0.320	0.0715	0.0175	0.00425	0.00108
0.8	2.36	1.69	1.30	1.04	0.850	0.715	0.600	0.500	0.425	0.364	0.0840	0.0212	0.00525	0.00140
0.9	2.45	1.75	1.38	1.11	0.920	0.775	0.650	0.550	0.475	0.404	0.0980	0.0260	0.00630	0.00165
1	2.54	1.85	1.45	1.18	0.975	0.825	0.700	0.595	0.510	0.444	0.113	0.0310	0.00840	0.00235
2	2.97	2.29	1.88	1.60	1.38	1.22	1.07	0.950	0.840	0.750	0.259	0.0950	0.0330	0.0115
3	3.20	2.50	2.10	1.82	1.60	1.42	1.28	1.15	1.05	0.960	0.388	0.165	0.0700	0.0275
4	3.36	2.66	2.25	1.97	1.75	1.58	1.42	1.30	1.20	1.10	0.495	0.235	0.112	0.0535
5	3.49	2.78	2.38	2.09	1.87	1.69	1.54	1.42	1.30	1.21	0.580	0.300	0.150	0.0715
6	3.59	2.90	2.47	2.18	1.95	1.78	1.65	1.52	1.40	1.30	0.660	0.360	0.195	0.0990
7	3.66	2.96	2.55	2.25	2.04	1.85	1.70	1.58	1.48	1.38	0.730	0.415	0.230	0.125
8	3.74	3.00	2.60	2.32	2.11	1.94	1.79	1.66	1.55	1.44	0.790	0.465	0.272	0.155
9	3.80	3.09	2.67	2.39	2.17	2.00	1.85	1.72	1.60	1.50	0.850	0.515	0.307	0.182
10	3.84	3.12	2.74	2.45	2.24	2.05	1.90	1.77	1.65	1.55	0.890	0.550	0.340	0.210

Note: For t' > 5, V(t',r') is about equal to $0.5W[(t'r')^2/4t]$, which is the well function in Table 2.2.2.
Source: From N.S. Boulton, 1954, The drawdown of water table under non-steady conditions near a pumped well in an unconfined formation, *Proc. Inst. Civ. Eng. (London)* 3, Pt. 2:564–579.

TABLE 2.2.5 CORRECTION FACTOR Ck

r'	0.03	0.04	0.06	0.08	0.1	0.2	0.4	0.6	0.8	1	2	4
C_k	−0.27	−0.24	−0.19	−0.16	−0.13	−0.05	0.02	0.05	0.05	0.05	0.03	0

$$\phi_w = \phi_o - \frac{Q}{2\pi K\phi_o}\left(m + \ell n \frac{\phi_o}{r_w}\right) \quad 2.2(29)$$

where m is a function of t' and can be obtained from a curve plotted through the following points:

t'	0.05	0.2	1	5
m	−0.043	0.087	0.512	1.288

—Y.S. Chae

References

Boulton, N.S. 1954. The drawdown of water table under non-steady conditions near a pumped well in an unconfined formation. Proc. Inst. Civ. Eng. (London) 3, pt 2:564–579.

Bouwer, H. 1978. *Groundwater hydrology.* McGraw-Hill, Inc.

Strack, O.D.L. 1989. *Groundwater mechanics.* Vol. 3, pt. 3:564–579. Prentice-Hall, Inc.

2.3
DETERMINING AQUIFER CHARACTERISTICS

Hydraulic conductivity K, transmissivity T, and storativity S are the hydraulic properties which characterize an aquifer. Before the quantities required to solve groundwater engineering problems, such as drawdown and rate of flow, can be calculated, the hydraulic properties of the aquifer K, S, and T must be determined.

Determining the hydraulic properties of an aquifer generally involves applying field data obtained from a pumping test. Other techniques such as auger-hole and piezometer methods can be used to determine K where the groundwater table or aquifers are shallow.

Pumping test technology is prominent in the evaluation of hydraulic properties. It involves observing the drawdown of the piezometric surface or water table in observation wells which are located some distance from the pumping well and have water pumped through them at a constant rate. Pumping test analysis applies the field data to some form of the Theis equation in general, such as

$$s = \frac{Q}{4\pi T} W(u, \alpha, \beta, \ldots) \quad 2.3(1)$$

where $u = Sr^2/4Tt$ and α,β = dimensionless factors defining particular aquifer system conditions. In general, matching the field data curve (usually a plot of s versus r^2/t) with the standard curve (known as the *type curve*) drawn between W and u for various control values of $\alpha, \beta, \ldots$, calculates the values of S and T. This process is explained in the next section. Techniques requiring no matching have since been developed.

Various site conditions are associated with a pumping test in a well–aquifer system. The following list summarizes different site conditions (Gupta 1989):

I. Type of pumping
   A. Drawdown
   B. Recovery
   C. Interference
II. State of flow
   A. Steady-state
   B. Nonsteady (transient) state
III. Area extent of aquifer
   A. Aquifer of infinite extent
   B. Aquifer bound by an impermeable boundary
   C. Aquifer bound by a recharge boundary
IV. Depth of well
   A. Fully penetrating well
   B. Partially penetrating well
V. Confined aquifer
   A. Nonleaky aquifer
   B. Leaky confining bed releasing water from storage
   C. Leaky confining bed not yielding water from storage but transmitting water from overlying layer
   D. Leaky aquifer in which the head in the overlying aquifer changes
VI. Unconfined aquifer
   A. Aquifer in which significant dewatering occurs
   B. Aquifer in which vertical flow occurs near the well
   C. Aquifer with delayed yield

Selecting a proper type curve is essential for the data analysis. During the last decades, several contributors have developed type curves for various site conditions or combinations of categories. Starting with Theis, who made the original type curve concept, other contributors to this field include Cooper and Jacob (1946) and Chow (1952) for confined aquifers, and Hantush and Jacob (1955), Neuman and Witherspoon (1969), Walton (1962), Boulton (1963) and Neuman (1972) for unconfined aquifers.

Confined Aquifers

This section discusses the methods used in determining aquifer characteristics for confined aquifers.

STEADY-STATE

The Thiem equation, Equation 9.3(12), gives the drawdown between two points (s_1 and s_2) measured at distances r_1 and r_2, respectively, as

$$s_1 - s_2 = s = \frac{Q}{2\pi T} \ell n \left(\frac{r_2}{r_1}\right) \qquad 2.3(2)$$

Hence, T can be calculated by

$$T = \frac{Q}{2\pi(s_1 - s_2)} \ell n \left(\frac{r_2}{r_1}\right) \qquad 2.3(3)$$

or from Figure 2.3.1, T can be obtained by

$$T = \frac{2.3Q}{2\pi} \frac{\Delta \log r}{\Delta s} \qquad 2.3(4)$$

Figure 2.2.2 shows that the drawdown between two points $s_1 - s_2$ reaches a constant value after a day or two. Therefore, Equation 2.3(3) can be used to determine T before the flow achieves a steady state.

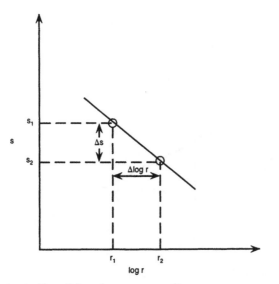

FIG. 2.3.1 Plot of drawdown s versus distance r.

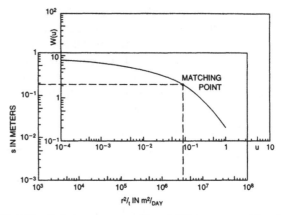

FIG. 2.3.2 Relations s versus r^2/t and W(u) versus u. (Reprinted from H. Bouwer, 1978, *Groundwater hydrology*, McGraw-Hill, Inc.)

Once T has been calculated, S can be determined with the transient-flow equations, Equations 2.2(14) and 2.2(20), as

$$W(u) = \frac{4\pi Ts}{Q} \longrightarrow T = \frac{Q}{4\pi s} W(u) \qquad 2.3(5)$$

$$u = \frac{r^2 S}{4Tt} \longrightarrow S = \frac{4Ttu}{r^2} \qquad 2.3(6)$$

Since T, Q, and s are known for a given r and t, W(u) can be obtained. With the use of Table 2.2.2, the corresponding value of u can be found. S can be calculated from Equation 2.3(6).

TRANSIENT-STATE

Three methods of analysis are the type-curve method (Theis), the Cooper–Jacob method, and the Chow method. These methods are briefly described.

Type Curve Method (Theis)

The Theis equation, Equations 2.2(20) and 2.2(14), can be written respectively as

$$\log s = \log \frac{Q}{4\pi T} + \log W(u) \qquad 2.3(7)$$

$$\log \frac{r^2}{t} = \log \frac{4T}{S} + \log u \qquad 2.3(8)$$

If these two equations are plotted on the same log–log paper, the resulting curves are the same shape but horizontally and vertically offset by the constants $Q/4\pi T$ and $4T/S$. If each curve is plotted on a separate sheet, the curves can be made to match when the sheets are overlapped as shown in Figure 2.3.2. An arbitrary point on the matching curve is selected, and the coordinates of this matching point are read horizontally and vertically on both graphs.

These values, s, r^2/t, u, and W(u) can then be used to calculate T and S from Equations 2.2(20) and 2.2(14).

The following example illustrates the Theis solution (H. Bouwer 1978). With the use of the drawdown data in Table 2.2.2, the data curve and type curve are overlapped to make the two curves match as shown in Figure 2.3.2. Four coordinates of the matching point are:

$$s = 0.167^m \qquad r^2/t = 3 \times 10^6 \ m^2/d$$

$$W(u) = 2.1 \qquad u = 8 \times 10^{-2} \qquad \text{2.3(9)}$$

Therefore,

$$T = \frac{Q}{4\pi s} W(u) = \frac{1000}{4\pi(0.167)}(2.1) = 1001 \ m^2/d \qquad \text{2.3(10)}$$

$$S = \frac{4Tu}{r^2/t} = \frac{4(1001)(8 \times 10^{-2})}{3 \times 10^6} = 0.0001 \qquad \text{2.3(11)}$$

Cooper-Jacob Method

Cooper and Jacob (1946) showed that when u becomes small (u << 1), the drawdown equation can be represented by Equation 2.2(22) as

$$s = \frac{2.3Q}{4\pi T} \log\left(\frac{2.25Tt}{Sr^2}\right) \qquad \text{2.3(12)}$$

On semilog paper, this equation represents a straight line with a slope of $2.3Q/4\pi T$. This equation can be plotted in three different ways: (1) s versus log t, (2) s versus log r, or (3) s versus log t/r^2 or log r^2/t.

DRAWDOWN–TIME ANALYSIS (s VERSUS log t)

The drawdown measurements s at a constant distance r are plotted against time as shown in Figure 2.3.3. The slope of the line is $2.3Q/4\pi T$ and is equal to

$$\frac{\Delta s}{\log \frac{t_2}{t_1}} = \frac{2.3Q}{4\pi T} \qquad \text{2.3(13)}$$

If a change in drawdown Δs is considered for one log cycle, then log $(t_2/t_1) = 1$, and this equation reduces to

$$\Delta s = \frac{2.3Q}{4\pi T} \qquad \text{2.3(14)}$$

or

$$T = \frac{2.3Q}{4\pi(\Delta s)} \qquad \text{2.3(15)}$$

When the straight line intersects the x axis, s = 0 and the time is t_o. Substituting these values in Equation 2.3(12) gives

$$0 = \frac{2.3Q}{4\pi T} \log \frac{2.25Tt_o}{r^2 S} \qquad \text{2.3(16)}$$

so

$$1 = \frac{2.25Tt_o}{r^2 S} \qquad \text{2.3(17)}$$

and

$$S = \frac{2.25Tt_o}{r^2} \qquad \text{2.3(18)}$$

Example: Figure 2.3.3 shows that $t_o = 1.6 \times 10^{-3}$ days and slope $\Delta s = 0.181$. These values yield T = 1011 m^2/d and S = 0.00009, which agree with the values for T and S obtained by the Theis solution.

DRAWDOWN–DISTANCE ANALYSIS (s VERSUS log r)

The drawdown measurements s are plotted against distance r at a given time t as shown in Figure 2.3.4. From similar considerations as in drawdown–time analysis

$$T = \frac{2.3Q}{2\pi(\Delta s)} \qquad \text{2.3(19)}$$

$$S = \frac{2.25Tt}{r_o^2} \qquad \text{2.3(20)}$$

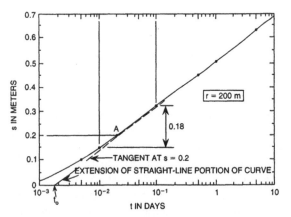

FIG. 2.3.3 Drawdown versus time plot. (Reprinted from H. Bouwer, 1978, *Groundwater hydrology*, McGraw-Hill, Inc.)

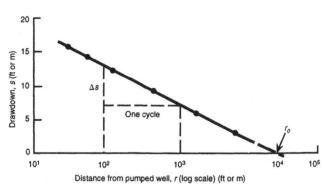

FIG. 2.3.4 Drawdown versus distance plot. (Reprinted from R.S. Gupta, 1989, *Hydrology and hydraulic systems,* Prentice-Hall, Inc.)

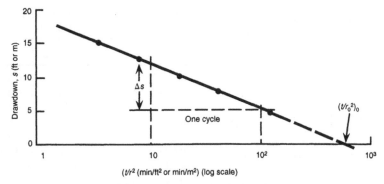

FIG. 2.3.5 Plot of drawdown versus combined time–distance. (Reprinted from R.S. Gupta, 1989, *Hydrology and hydraulic systems,* Prentice-Hall, Inc.)

DRAWDOWN–COMBINED-TIME–DISTANCE ANALYSIS (s VERSUS log r²/t)

The drawdown measurements in many wells at various times are plotted as shown in Figure 2.3.5. Similarly as before

$$T = \frac{2.3Q}{4\pi(\Delta s)} \qquad 2.3(21)$$

$$S = 2.25T\left(\frac{t}{r^2}\right)_0 \qquad 2.3(22)$$

Chow Method

Chow's procedure (1952) combines the approach of Theis and Cooper–Jacob and introduces the function

$$F(u) = \frac{W(u)e^u}{2.3} = \frac{s}{\Delta s/\log(t_2/t_1)} \qquad 2.3(23)$$

where s is the drawdown at a point. The relation between F(u), W(u), and u is shown in Figure 2.3.6. For one log cycle on a time scale

$$\log(t_2/t_1) = 1 \qquad 2.3(24)$$

and

$$F(u) = \frac{s}{\Delta s} \qquad 2.3(25)$$

From the drawdown–time curve, obtain s at an arbitrary point and Δs over one log cycle. The ratio s/Δs is equal to F(u) in the Equation 2.3(25). F(u), W(u), and u can be obtained from Figure 2.3.6. With W(u), u, s, and t known, T and S can be calculated with Equations 2.2(20) and 2.2(14).

Example: Point A in Figure 2.3.3 gives s = 0.2^m and Δs = 0.18^m at r = 200^m and F(u) = 0.2/0.18 = 1.11. From Figure 2.3.6, W(u) = 2.2 and u = 0.065. Substituting into Equations 2.2(20) and 2.2(14) yields T = 875 m²/d and S = 0.00011, which reasonably agree with the values obtained by the two methods just described.

Recovery Test

Figure 2.3.7 schematically shows a recovery test in which the water level in the observation wells rises when pumping stops after the pumping test is complete. Since the principle of superposition applies, the drawdown s' after the pumping test is complete can be expressed as

$$s' = \frac{Q}{4\pi T}\,\ell n\left(\frac{2.25Tt}{r^2S}\right) - \frac{Q-Q'}{4\pi T}\,\ell n\left(\frac{2.25Tt'}{r^2S}\right) \qquad 2.3(26)$$

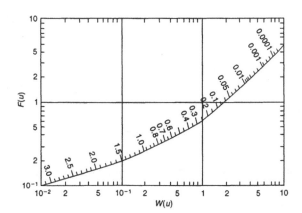

FIG. 2.3.6 Relations between F(u), W(u), and u. (Reprinted from V.T. Chow, 1952, On the determination of transmissivity and storage coefficients from pumping test data, *Trans. Am. Geoph. Union* 33:397–404.)

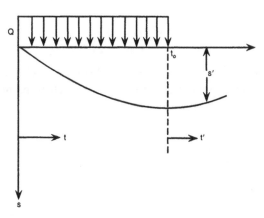

FIG. 2.3.7 Recovery test.

where Q' is the rate of flow, and t' is the time after the pumping stops, respectively. Since $Q' = 0$, s' becomes

$$s' = \frac{Q}{4\pi T}\ell n\,\frac{t}{t'} = \frac{2.3Q}{4\pi T}\log\frac{t}{t'} \qquad 2.3(27)$$

Thus, T can be calculated as

$$T = \frac{2.3Q}{4\pi\Delta s'} \qquad 2.3(28)$$

However, S cannot be determined from the recovery test.

Semiconfined (Leaky) Aquifers

This section discusses the methods used in determining aquifer characteristics for semiconfined (leaky) aquifers (Bouwer 1978).

STEADY-STATE

The DeGlee–Hantush–Jacob method (DeGlee 1930, 1951; Hantush and Jacob 1955) and the Hantush method (1956, 1964) are used to determine the aquifer characteristics in semiconfined aquifers under steady-state conditions.

De Glee–Hantush–Jacob Method

The drawdown in a semiconfined aquifer is given by Equation 9.3(18) as

$$s = \frac{Q}{2\pi T}K_o\!\left(\frac{r}{\lambda}\right) \qquad 2.3(29)$$

where $K_o(r/\lambda)$ = modified Bessel function of zero order and second kind, and $\lambda = \sqrt{Tc}$ as defined before. The values of $K_o(r/\lambda)$ versus r/λ are shown in Table 2.3.1. The value T can be determined as in a confined aquifer with the use of the matching procedure. The data curve is obtained from a plot of s versus r on log–log paper, and the type curve is obtained from a plot of $K_o(r/\lambda)$ versus r/λ. Overlapping these two plots matches the two curves, and four coordinates of an arbitrary selected print on the matching curve are noted. The value T is then calculated from Equation 2.3(29) as

$$T = \frac{Q}{2\pi s}K_o\!\left(\frac{r}{\lambda}\right) \qquad 2.3(30)$$

The resistance c can be determined from $c = \lambda^2/T$ when T and the values of r and r/λ of the matching point are substituted into this equation.

Hantush Method

Equation 1.3(19) shows that when $r/\lambda \ll 1$, the drawdown can be approximated by

$$s = \frac{2.3Q}{2\pi T}\log\frac{1.123\lambda}{r} \qquad 2.3(31)$$

TABLE 2.3.1   VALUES OF THE FUNCTIONS $K_0(X)$ AND EXP $(X)K_0(X)$

x	$K_0(x)$	$exp\,(x)K_0(x)$	x	$K_0(x)$	$exp\,(x)K_0(x)$
0.01	4.72	4.77	0.35	1.23	1.75
0.015	4.32	4.38	0.40	1.11	1.6
0.02	4.03	4.11	0.45	1.01	1.59
0.025	3.81	3.91	0.50	0.92	1.52
0.03	3.62	3.73	0.55	0.85	1.47
0.035	3.47	3.59	0.60	0.78	1.42
0.04	3.34	3.47	0.65	0.72	1.37
0.045	3.22	3.37	0.70	0.66	1.33
0.05	3.11	3.27	0.75	0.61	1.29
0.055	3.02	3.19	0.80	0.57	1.26
0.06	2.93	3.11	0.85	0.52	1.23
0.065	2.85	3.05	0.90	0.49	1.20
0.07	2.78	2.98	0.95	0.45	1.17
0.075	2.71	2.92	1.0	0.42	1.14
0.08	2.65	2.87	1.5	0.21	0.96
0.085	2.59	2.82	2.0	0.11	0.84
0.09	2.53	2.77	2.5	0.062	0.760
0.095	2.48	2.72	3.0	0.035	0.698
0.10	2.43	2.68	3.5	0.020	0.649
0.15	2.03	2.36	4.0	0.011	0.609
0.20	1.75	2.14	4.5	0.006	0.576
0.25	1.54	1.98	5.0	0.004	0.548
0.30	1.37	1.85			

Source: Adapted from M.S. Hantush, 1956, Analysis of data from pumping tests in leaky aquifers, *Transactions American Geophysical Union* 37:702–14 and C.W. Fetter, 1988, *Applied hydrogeology*, 2d ed., Macmillan.

A plot of s versus log r forms a straight line, the slope of which is $2.3Q/2\pi T$. If Δs is taken over one log cycle, then T can be calculated as

$$T = \frac{2.3Q}{2\pi(\Delta s)} \qquad 2.3(32)$$

Extending the straight line into the abscissa yields the intercept r_o where $s = 0$. Then, from Equation 2.3(31)

$$0 = \log\frac{1.123\lambda}{r_o} \qquad 2.3(33)$$

so

$$\lambda = r_o/1.123 \qquad 2.3(34)$$

and

$$c = \frac{\lambda^2}{T} = \frac{r_o^2}{1.25T} \qquad 2.3(35)$$

Note that this method does not require the matching procedure.

TRANSIENT-STATE

Hantush and Jacob (1955) showed that the drawdown in a semiconfined aquifer is described by

$$s = \frac{Q}{4\pi T} W\left(u, \frac{r}{\lambda}\right) \qquad 2.3(36)$$

where

$$u = \frac{r^2 S}{4Tt} \qquad 2.3(37)$$

Equation 2.3(36) is similar to Equation 2.2(20) for a confined aquifer except that the well function contains the additional term r/λ. The values of $W(u, r/\lambda)$ are given in Table 2.3.2.

Walton Method

Walton's solution (1962) of Equation 2.3(36) is similar to the Theis method for a confined aquifer. Plotting s versus t/r^2 gives the data curve. Plotting $W(u, r/\lambda)$ versus u for various values of r/λ gives several type curves. Figure 2.3.8 shows the type curves. The data curve is superimposed on the type curves to get the best fitting curve. Again, four coordinates of a match point are read on both graphs. The resulting values of $W(u, r/\lambda)$ and s are substituted into Equation 2.3(36) to calculate T. The value of S is obtained from Equation 2.3(37) when u, t/r^2, and T are substituted. The value c is calculated from $c = \lambda^2/T$ where λ is obtained from the r/λ value of the best fitting curve.

Hantush's Inflection Point Method

Hantush's procedure (1956) for calculating T, S, and c from pumping test data utilizes the halfway point or inflection point on a curve relating s to $\log t$. The inflection point is the point where the drawdown s is one-half the final or equilibrium drawdown as

$$s = \frac{Q}{4\pi T} K_o\left(\frac{r}{\lambda}\right) \qquad 2.3(38)$$

The value u at the inflection point is

$$\frac{r}{2\lambda} = u = \frac{r^2 S}{4Tt_i} \qquad 2.3(39)$$

where t_i is t at the inflection point. The ratio between the drawdown and the slope of the curve at the inflection point Δs expressed as the drawdown per unit log cycle of t is derived as

$$2.3 \frac{s}{\Delta s} = e^{r/\lambda} \cdot K_o\left(\frac{r}{\lambda}\right) \qquad 2.3(40)$$

The values of function $e^{r/\lambda} \cdot K_o(r/\lambda)$ versus r/λ are in Table 2.3.1.

To determine T, S, and c from pumping test data, follow the following procedure:

1. Plot drawdown–time on semilog paper (s–$\log t$).
2. Locate the inflection point P where $s = 1/2 \times$ final drawdown.
3. Draw a line tangent to the curve at point P, and determine the corresponding value of t_i and the slope Δs.
4. Substitute s and Δs values into Equation 2.3(40) to obtain $e^{r/\lambda} \cdot K_o(r/\lambda)$, and determine the corresponding value of r/λ and $K_o r/\lambda$ from Table 2.3.1.
5. Determine T from Equation 2.3(38).
6. Determine S from Equation 2.3(39).
7. Determine c from $c = \lambda^2/T$.

Unconfined Aquifers

This section discusses the methods used in determining aquifer characteristics for unconfined aquifers.

STEADY-STATE

As previously explained, the equation of groundwater flow for unconfined aquifers reduces to the same form as that for confined aquifers except that the thickness of the aquifer is not constant but varies as the aquifer is dewatered. Therefore, the flow must be expressed through an average thickness of the aquifer ϕ_{av}. The Thiem equation is then

$$Q = \frac{\pi K(\phi_2^2 - \phi_1^2)}{\ln\left(\frac{r_2}{r_1}\right)} = \frac{\pi K 2\phi_{av}(\phi_2 - \phi_1)}{\ln\left(\frac{r_2}{r_1}\right)}$$
$$= \frac{2\pi T_{av}(\phi_2 - \phi_1)}{\ln\left(\frac{r_2}{r_1}\right)} \qquad 2.3(41)$$

where $\phi_2 = \phi_o - s_2$ and $\phi_1 = \phi_o - s_1$.
From Equation 2.3(41),

$$T_{av} = \frac{Q \ln\left(\frac{r_2}{r_1}\right)}{2\pi \cdot (s_1 - s_2)} \qquad 2.3(42)$$

which is the same form as that for a confined aquifer. The transmissibility of the aquifer T is then

$$T = \frac{2\phi_o}{2\phi_o - s_1 - s_2} \cdot T_{av} \qquad 2.3(43)$$

Once T has been determined, S can be obtained in the same manner as a confined aquifer. Note that when the steady-state method is applied, pumping does not have to continue until true steady-state conditions are reached since $\Delta s = s_1 - s_2$ reaches an essentially constant value after a few days of pumping.

TRANSIENT-STATE

As explained previously, the transient flow of groundwater in an unconfined aquifer occurs from two types of storage: phreatic and elastic. As water is pumped out of the aquifer, the decline in pressure in the aquifer yields water due to the elastic storage of the aquifer storativity S_e, and the declining water table also yields water as it drains under gravity. Unlike the confined aquifer, the release of wa-

TABLE 2.3.2 VALUES OF W(U,R/Λ) FOR DIFFERENT VALUES OF U AND R/Λ

u ＼ r/λ	0.002	0.004	0.006	0.008	0.01	0.02	0.04	0.06	0.08	0.1	0.2	0.4	0.6	0.8	1	2	4	6	8
0	12.7	11.3	10.5	9.89	9.44	8.06	6.67	5.87	5.29	4.85	3.51	2.23	1.55	1.13	0.842	0.228	0.0223	0.0025	0.0003
0.000002	12.1	11.2	10.5	9.89	9.44	8.06	6.67	5.87	5.29	4.85	3.51	2.23	1.55	1.13	0.842	0.228	0.0223	0.0025	0.0003
4	11.6	11.1	10.4	9.88	9.44	8.06	6.67	5.87	5.29	4.85	3.51	2.23	1.55	1.13	0.842	0.228	0.0223	0.0025	0.0003
6	11.3	10.9	10.4	9.87	9.44	8.06	6.67	5.87	5.29	4.85	3.51	2.23	1.55	1.13	0.842	0.228	0.0223	0.0025	0.0003
8	11.0	10.7	10.3	9.84	9.43	8.06	6.67	5.87	5.29	4.85	3.51	2.23	1.55	1.13	0.842	0.228	0.0223	0.0025	0.0003
0.00001	10.8	10.6	10.2	9.80	9.42	8.06	6.67	5.87	5.29	4.85	3.51	2.23	1.55	1.13	0.842	0.228	0.0223	0.0025	0.0003
2	10.2	10.1	9.84	9.58	9.30	8.06	6.67	5.87	5.29	4.85	3.51	2.23	1.55	1.13	0.842	0.228	0.0223	0.0025	0.0003
4	9.52	9.45	9.34	9.19	9.01	8.03	6.67	5.87	5.29	4.85	3.51	2.23	1.55	1.13	0.842	0.228	0.0223	0.0025	0.0003
6	9.13	9.08	9.00	8.89	8.77	7.98	6.67	5.87	5.29	4.85	3.51	2.23	1.55	1.13	0.842	0.228	0.0223	0.0025	0.0003
8	8.84	8.81	8.75	8.67	8.57	7.91	6.67	5.87	5.29	4.85	3.51	2.23	1.55	1.13	0.842	0.228	0.0223	0.0025	0.0003
0.0001	8.62	8.59	8.55	8.48	8.40	7.84	6.67	5.87	5.29	4.85	3.51	2.23	1.55	1.13	0.842	0.228	0.0223	0.0025	0.0003
2	7.94	7.92	7.90	7.86	7.82	7.50	6.62	5.86	5.29	4.85	3.51	2.23	1.55	1.13	0.842	0.228	0.0223	0.0025	0.0003
4	7.24	7.24	7.22	7.21	7.19	7.01	6.45	5.83	5.29	4.85	3.51	2.23	1.55	1.13	0.842	0.228	0.0223	0.0025	0.0003
6	6.84	6.84	6.84	6.82	6.80	6.68	6.27	5.77	5.27	4.85	3.51	2.23	1.55	1.13	0.842	0.228	0.0223	0.0025	0.0003
8	6.55	6.55	6.55	6.53	6.52	6.43	6.11	5.69	5.25	4.85	3.51	2.23	1.55	1.13	0.842	0.228	0.0223	0.0025	0.0003
0.001	6.33	6.33	6.33	6.32	6.31	6.23	5.97	5.61	5.21	4.83	3.51	2.23	1.55	1.13	0.842	0.228	0.0223	0.0025	0.0003
2	5.64	5.64	5.64	5.63	5.63	5.59	5.45	5.24	4.98	4.71	3.50	2.23	1.55	1.13	0.842	0.228	0.0223	0.0025	0.0003
4	4.95	4.95	4.95	4.94	4.94	4.92	4.85	4.74	4.59	4.42	3.48	2.23	1.55	1.13	0.842	0.228	0.0223	0.0025	0.0003
6	4.54	4.54	4.54	4.54	4.54	4.53	4.48	4.41	4.30	4.18	3.43	2.23	1.55	1.13	0.842	0.228	0.0223	0.0025	0.0003
8	4.26	4.26	4.26	4.26	4.26	4.25	4.21	4.15	4.08	3.98	3.36	2.23	1.55	1.13	0.842	0.228	0.0223	0.0025	0.0003
0.01	4.04	4.04	4.04	4.04	4.04	4.03	4.00	3.95	3.89	3.81	3.29	2.23	1.55	1.13	0.842	0.228	0.0223	0.0025	0.0003
2	3.35	3.35	3.35	3.35	3.35	3.35	3.34	3.31	3.28	3.24	2.95	2.18	1.55	1.13	0.839	0.228	0.0223	0.0025	0.0003
4	2.68	2.68	2.68	2.68	2.68	2.68	2.67	2.66	2.65	2.63	2.48	2.02	1.52	1.13	0.832	0.228	0.0223	0.0025	0.0003
6	2.30	2.30	2.30	2.30	2.30	2.30	2.29	2.28	2.27	2.26	2.17	1.85	1.46	1.11	0.839	0.228	0.0223	0.0025	0.0003
8	2.03	2.03	2.03	2.03	2.03	2.03	2.02	2.02	2.01	2.00	1.94	1.69	1.39	1.08	0.832	0.228	0.0223	0.0025	0.0003
0.1	1.82	1.82	1.82	1.82	1.82	1.82	1.82	1.82	1.81	1.80	1.75	1.56	1.31	1.05	0.819	0.228	0.0223	0.0025	0.0003
2	1.22	1.22	1.22	1.22	1.22	1.22	1.22	1.22	1.22	1.22	1.19	1.11	0.996	0.857	0.715	0.227	0.0223	0.0025	0.0003
4	0.702	0.702	0.702	0.702	0.702	0.702	0.702	0.702	0.701	0.700	0.693	0.665	0.621	0.565	0.502	0.210	0.0223	0.0025	0.0003
6	0.454	0.454	0.454	0.454	0.454	0.454	0.454	0.454	0.454	0.453	0.450	0.436	0.415	0.387	0.354	0.177	0.0222	0.0025	0.0003
8	0.311	0.311	0.311	0.311	0.311	0.311	0.311	0.310	0.310	0.310	0.308	0.301	0.289	0.273	0.254	0.144	0.0218	0.0025	0.0003
1	0.219	0.219	0.219	0.219	0.219	0.219	0.219	0.219	0.219	0.219	0.218	0.213	0.206	0.197	0.185	0.114	0.0207	0.0025	0.0003
2	0.049	0.049	0.049	0.049	0.049	0.049	0.049	0.049	0.049	0.049	0.049	0.048	0.047	0.046	0.044	0.034	0.011	0.0021	0.0002
4	0.0038	0.0038	0.0038	0.0038	0.0038	0.0038	0.0038	0.0038	0.0038	0.0038	0.0038	0.0038	0.0037	0.0037	0.0036	0.0031	0.0016	0.0006	0
6	0.0004	0.0004	0.0004	0.0004	0.0004	0.0004	0.0004	0.0004	0.0004	0.0004	0.0004	0.0004	0.0004	0.0004	0.0004	0.0003	0.0002	0.0001	0
8	0	0	0	0	0	0	0	0	0	0	0	0	0	0	0	0	0	0	0

Source: From M.S. Hantush, 1956, Analysis of data from pumping tests in leaky aquifers, *Transactions American Geophysical Union* 37:702–14. Reference to the original article is made for more extensive tables and expression of W(u,r/λ) in more significant figures (See also M.S. Hantush, 1964, Hydraulics of wells, In Vol. 1 of *Advances in hydroscience*, edited by V.T. Chow [New York and London: Academic Press]:281–432) and H. Bouwer, 1978, *Groundwater hydrology*, McGraw-Hill, Inc.

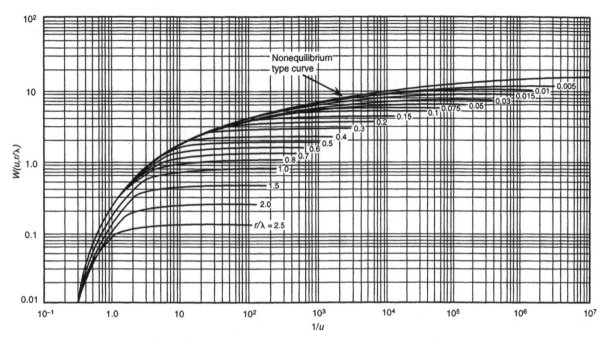

FIG. 2.3.8 Type curves for a leaky aquifer. (Reprinted from C.W. Fetter, 1988, *Applied hydrogeology*, 2d ed., Macmillan Pub. Co.)

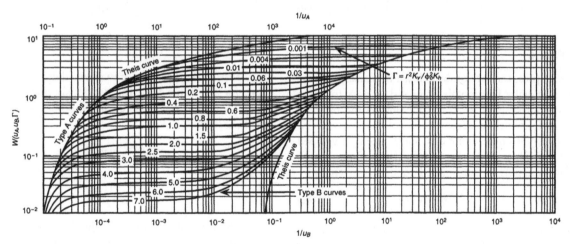

FIG. 2.3.9 Type curves and curves for a delayed yield. (Reprinted from C.W. Fetter, 1988, *Applied hydrogeology*, 2d ed., Macmillan Pub. Co.)

ter from storage is not immediate in response to the drop of the water table. The yield is delayed depending on the elastic and phreatic storativity of the aquifer. Accordingly, the delayed yield produces a sigmoid drawdown curve as shown in Figure 2.3.9.

Essentially, three distinct phases of drawdown–time (s–t) relations occur as shown in Figure 2.3.9: initial phase, intermediate phase, and final phase.

Initial Phase

As the pumping begins, a small amount of water is released from the aquifer under the pressure drop due to the

compression of the aquifer. During this stage, the aquifer behaves as a confined aquifer, and the flow is essentially horizontal. The drawdown–time data follow a Theis-type curve (type A) for elastic storativity S_e, which is small.

Intermediate Phase

Following the initial phase, as the water table begins to decline, water is drawn primarily from the gravity drainage of the aquifer. The flow at this stage has both horizontal and vertical components, and the s–t relationship is a function of the ratio of the horizontal to vertical hydraulic con-

TABLE 2.3.3 VALUES OF THE FUNCTION $W(u_A, \Gamma)$ FOR WATER TABLE AQUIFERS

$1/u_A$	$\Gamma = 0.001$	$\Gamma = 0.01$	$\Gamma = 0.06$	$\Gamma = 0.2$	$\Gamma = 0.6$	$\Gamma = 1.0$	$\Gamma = 2.0$	$\Gamma = 4.0$	$\Gamma = 6.0$
4.0×10^{-1}	2.48×10^{-2}	2.41×10^{-2}	2.30×10^{-2}	2.14×10^{-2}	1.88×10^{-2}	1.70×10^{-2}	1.38×10^{-2}	9.33×10^{-3}	6.39×10^{-3}
8.0×10^{-1}	1.45×10^{-1}	1.40×10^{-1}	1.31×10^{-1}	1.19×10^{-1}	9.88×10^{-2}	8.49×10^{-2}	6.03×10^{-2}	3.17×10^{-2}	1.74×10^{-2}
1.4×10^{0}	3.58×10^{-1}	3.45×10^{-1}	3.18×10^{-1}	2.79×10^{-1}	2.17×10^{-1}	1.75×10^{-1}	1.07×10^{-1}	4.45×10^{-2}	2.10×10^{-2}
2.4×10^{0}	6.62×10^{-1}	6.33×10^{-1}	5.70×10^{-1}	4.83×10^{-1}	3.43×10^{-1}	2.56×10^{-1}	1.33×10^{-1}	4.76×10^{-2}	2.14×10^{-2}
4.0×10^{0}	1.02×10^{0}	9.63×10^{-1}	8.49×10^{-1}	6.88×10^{-1}	4.38×10^{-1}	3.00×10^{-1}	1.40×10^{-1}	4.78×10^{-2}	2.15×10^{-2}
8.0×10^{0}	1.57×10^{0}	1.46×10^{0}	1.23×10^{0}	9.18×10^{-1}	4.97×10^{-1}	3.17×10^{-1}	1.41×10^{-1}		
1.4×10^{1}	2.05×10^{0}	1.88×10^{0}	1.51×10^{0}	1.03×10^{0}	5.07×10^{-1}				
2.4×10^{1}	2.52×10^{0}	2.27×10^{0}	1.73×10^{0}	1.07×10^{0}					
4.0×10^{1}	2.97×10^{0}	2.61×10^{0}	1.85×10^{0}	1.08×10^{0}					
8.0×10^{1}	3.56×10^{0}	3.00×10^{0}	1.92×10^{0}						
1.4×10^{2}	4.01×10^{0}	3.23×10^{0}	1.93×10^{0}						
2.4×10^{2}	4.42×10^{0}	3.37×10^{0}	1.94×10^{0}						
4.0×10^{2}	4.77×10^{0}	3.43×10^{0}							
8.0×10^{2}	5.16×10^{0}	3.45×10^{0}							
1.4×10^{3}	5.40×10^{0}	3.46×10^{0}							
2.4×10^{3}	5.54×10^{0}								
4.0×10^{3}	5.59×10^{0}								
8.0×10^{3}	5.62×10^{0}								
1.4×10^{4}	5.62×10^{0}	3.46×10^{0}	1.94×10^{0}	1.08×10^{0}	5.07×10^{-1}	3.17×10^{-1}	1.41×10^{-1}	4.78×10^{-2}	2.15×10^{-2}

Source: Adapted from S.P. Neuman, 1975, *Water Resources Research* 11:329–42 and C.W. Fetter, 1988, *Applied hydrogeology*, 2d ed., Macmillan.

ductivity of the aquifer, the distance to the pumping well, and the aquifer thickness.

Final Phase

As time elapses, the rate of drawdown decreases, and the flow is essentially horizontal. The s–t data now follow a Theis-type curve (type B) corresponding to the phreatic storativity S_p, which is large.

Several type-curve solutions have been developed (Walton 1962), such as the one shown in Figure 2.3.9. The flow equation for unconfined aquifers is given by

$$s = \frac{Q}{4\pi T} W(u_A, u_B, \Gamma) \qquad 2.3(44)$$

where $W(u_A, u_B, \Gamma)$ is the well function, and

$$u_A = \frac{S_e r^2}{4Tt} \quad \text{(for early drawdown data)} \qquad 2.3(45)$$

$$u_B = \frac{S_p r^2}{4Tt} \quad \text{(for later drawdown data)} \qquad 2.3(46)$$

and

$$\Gamma = \frac{r^2 K_v}{\phi_o^2 K_h} \qquad 2.3(47)$$

The values of $W(u_A, \Gamma)$ and $W(u_B, \Gamma)$ are given in Tables 2.3.3 and 2.3.4. The type curves are used to evaluate the field data for drawdown and time with the use of the following procedure (Fetter 1988):

1. Superpose the late drawdown–time data on the type-B curves for the best fit. At any match point, determine the values of $W(u_B, \Gamma)$, u_B, t, and s. Obtain the value Γ from the type curve. Calculate T and S_p from

$$T = \frac{Q}{4\pi s} W(u_B, \Gamma) \qquad 2.3(48)$$

$$S_p = \frac{4Ttu_B}{r^2} \qquad 2.3(49)$$

2. Superpose the early drawdown data on the type-A curve for the value Γ of the previously matched type-B curve. Determine a new set of match points, and calculate T and S_e from

$$T = \frac{Q}{4\pi s} W(u_A, \Gamma) \qquad 2.3(50)$$

$$S_e = \frac{4Ttu_A}{r^2} \qquad 2.3(51)$$

The calculated value of T should be approximately equal to that computed from the type-B curve.

3. Determine K_h and K_v from

$$K_h = \frac{T}{\phi_o} \qquad 2.3(52)$$

$$K_v = \frac{\Gamma \phi_o^2 K_h}{r^2} \qquad 2.3(53)$$

Slug Tests

A slug test is a simple and inexpensive way of determining local values of aquifer properties. Instead of the well being pumped for a period of time, a volume of water is suddenly removed or added to the well casing, and recovery or drawdown are observed over time. Through careful evaluation of the drawdown curve and knowledge of the well screen geometry, the hydraulic conductivity of an aquifer can be derived (Bedient 1994).

TABLE 2.3.4   VALUES OF THE FUNCTION W(U$_B$,Γ) FOR WATER TABLE AQUIFERS

$1/u_B$	Γ = 0.001	Γ = 0.01	Γ = 0.06	Γ = 0.2	Γ = 0.6	Γ = 1.0	Γ = 2.0	Γ = 4.0	Γ = 6.0
4.0×10^{-4}	5.62×10^0	3.46×10^0	1.94×10^0	1.09×10^0	5.08×10^{-1}	3.18×10^{-1}	1.42×10^{-1}	4.79×10^{-2}	2.15×10^{-2}
8.0×10^{-4}								4.80×10^{-2}	2.16×10^{-2}
1.4×10^{-3}								4.81×10^{-2}	2.17×10^{-2}
2.4×10^{-3}								4.84×10^{-2}	2.19×10^{-2}
4.0×10^{-3}					5.08×10^{-1}	3.18×10^{-1}	1.42×10^{-1}	4.88×10^{-2}	2.21×10^{-2}
8.0×10^{-3}					5.09×10^{-1}	3.19×10^{-1}	1.43×10^{-1}	4.96×10^{-2}	2.28×10^{-2}
1.4×10^{-2}					5.10×10^{-1}	3.21×10^{-1}	1.45×10^{-1}	5.09×10^{-2}	2.39×10^{-2}
2.4×10^{-2}					5.12×10^{-1}	3.23×10^{-1}	1.47×10^{-1}	5.32×10^{-2}	2.57×10^{-2}
4.0×10^{-2}					5.16×10^{-1}	3.27×10^{-1}	1.52×10^{-1}	5.68×10^{-2}	2.86×10^{-2}
8.0×10^{-2}				1.09×10^0	5.24×10^{-1}	3.37×10^{-1}	1.62×10^{-1}	6.61×10^{-2}	3.62×10^{-2}
1.4×10^{-1}			1.94×10^0	1.10×10^0	5.37×10^{-1}	3.50×10^{-1}	1.78×10^{-1}	8.06×10^{-2}	4.86×10^{-2}
2.4×10^{-1}			1.95×10^0	1.11×10^0	5.57×10^{-1}	3.74×10^{-1}	2.05×10^{-1}	1.06×10^{-1}	7.14×10^{-2}
4.0×10^{-1}			1.96×10^0	1.13×10^0	5.89×10^{-1}	4.12×10^{-1}	2.48×10^{-1}	1.49×10^{-1}	1.13×10^{-1}
8.0×10^{-1}	5.62×10^0	3.46×10^0	1.98×10^0	1.18×10^0	6.67×10^{-1}	5.06×10^{-1}	3.57×10^{-1}	2.66×10^{-1}	2.31×10^{-1}
1.4×10^0	5.63×10^0	3.47×10^0	2.01×10^0	1.24×10^0	7.80×10^{-1}	6.42×10^{-1}	5.17×10^{-1}	4.45×10^{-1}	4.19×10^{-1}
2.4×10^0	5.63×10^0	3.49×10^0	2.06×10^0	1.35×10^0	9.54×10^{-1}	8.50×10^{-1}	7.63×10^{-1}	7.18×10^{-1}	7.03×10^{-1}
4.0×10^0	5.63×10^0	3.51×10^0	2.13×10^0	1.50×10^0	1.20×10^0	1.13×10^0	1.08×10^0	1.06×10^0	1.05×10^0
8.0×10^0	5.64×10^0	3.56×10^0	2.31×10^0	1.85×10^0	1.68×10^0	1.65×10^0	1.63×10^0	1.63×10^0	1.63×10^0
1.4×10^1	5.65×10^0	3.63×10^0	2.55×10^0	2.23×10^0	2.15×10^0	2.14×10^0	2.14×10^0	2.14×10^0	2.14×10^0
2.4×10^1	5.67×10^0	3.74×10^0	2.86×10^0	2.68×10^0	2.65×10^0	2.65×10^0	2.64×10^0	2.64×10^0	2.64×10^0
4.0×10^1	5.70×10^0	3.90×10^0	3.24×10^0	3.15×10^0	3.14×10^0	3.14×10^0	3.14×10^0	3.14×10^0	3.14×10^0
8.0×10^1	5.76×10^0	4.22×10^0	3.85×10^0	3.82×10^0	3.82×10^0	3.82×10^0	3.82×10^0	3.82×10^0	3.82×10^0
1.4×10^2	5.85×10^0	4.58×10^0	4.38×10^0	4.37×10^0	4.37×10^0	4.37×10^0	4.37×10^0	4.37×10^0	4.37×10^0
2.4×10^2	5.99×10^0	5.00×10^0	4.91×10^0	4.91×10^0	4.91×10^0	4.91×10^0	4.91×10^0	4.91×10^0	4.91×10^0
4.0×10^2	6.16×10^0	5.46×10^0	5.42×10^0	5.42×10^0	5.42×10^0	5.42×10^0	5.42×10^0	5.42×10^0	5.42×10^0
8.0×10^2	6.47×10^0	6.11×10^0	6.11×10^0	6.11×10^0	6.11×10^0	6.11×10^0	6.11×10^0	6.11×10^0	6.11×10^0
1.4×10^3	6.67×10^0	6.67×10^0	6.67×10^0	6.67×10^0	6.67×10^0	6.67×10^0	6.67×10^0	6.67×10^0	6.67×10^0
2.4×10^3	7.21×10^0	7.21×10^0	7.21×10^0	7.21×10^0	7.21×10^0	7.21×10^0	7.21×10^0	7.21×10^0	7.21×10^0
4.0×10^3	7.72×10^0	7.72×10^0	7.72×10^0	7.72×10^0	7.72×10^0	7.72×10^0	7.72×10^0	7.72×10^0	7.72×10^0
8.0×10^3	8.41×10^0	8.41×10^0	8.41×10^0	8.41×10^0	8.41×10^0	8.41×10^0	8.41×10^0	8.41×10^0	8.41×10^0
1.4×10^4	8.97×10^0	8.97×10^0	8.97×10^0	8.97×10^0	8.97×10^0	8.97×10^0	8.97×10^0	8.97×10^0	8.97×10^0
2.4×10^4	9.51×10^0	9.51×10^0	9.51×10^0	9.51×10^0	9.51×10^0	9.51×10^0	9.51×10^0	9.51×10^0	9.51×10^0
4.0×10^4	1.94×10^1	1.94×10^1	1.94×10^1	1.94×10^1	1.94×10^1	1.94×10^1	1.94×10^1	1.94×10^1	1.94×10^1

Source: Adapted from S.P. Neuman, 1975, *Water Resources Research* 11:329–42 and C.W. Fetter, 1988, *Applied hydrogeology*, 2d ed., Macmillan.

Hvorslev (1951) developed the simplest slug test method in a piezometer, which relates the flow rate Q(t) at the piezometer at any time to the hydraulic conductivity and unrecovered head distance $H_o - h$ in Figure 2.3.10 by

$$Q(t) = \pi r^2 \frac{dh}{dt} = FK(H_o - h) \qquad 2.3(54)$$

where F is a factor that depends on the shape and the dimensions of the piezometer intake. If $Q = Q_o$ at $t = 0$, then Q(t) decreases toward zero as time increases. Hvorslev defined the basic time lag $T_o = \pi r^2/FK$ and solved Equation 2.3(54) with initial conditions $h = H_o$ at $t = 0$. Thus

$$\frac{H - h}{H - H_o} = e^{-t/T_o} \qquad 2.3(55)$$

When recovery $H - h/H - H_o$ versus time is plotted on semilog paper, T_o is noted at t where recovery equals 37% of the initial change. For the piezometer intake length divided by radius, L/R greater than 8, Hvorslev has evaluated the shape factor F and obtained an equation for K as

$$K = \frac{r^2 \ell n(L/R)}{2LT_o} \qquad 2.3(56)$$

Several other slug test methods have been developed for confined aquifers by Cooper et al. (1967) and Papadopoulos et al. (1973). These methods are similar to Theis's in which a curve-matching procedure is used to obtain S and T for a given aquifer. Figure 2.3.11 shows the slug test curves developed by Papadopoulos for various values of variable α, defined as

$$\alpha = \frac{r_s^2}{r_c^2} S \qquad 2.3(57)$$

The obtained data are plotted and matched to the plotted type curves for a best match, from which α is selected for a particular curve. The vertical time axis t which overlays the vertical axis for $Tt/r_c^2 = 1.0$ is selected, and a value of T can then be found from $T = 1.0 r_s^2/t_1$. Then, the value of S can be found from the definition of α. The method is representative of the formation only in the immediate vicinity of the test hole and should be used with caution (Bedient 1994).

The most commonly used method for determining hydraulic conductivity in groundwater investigation is the Bouwer and Rice (1976) slug test shown in Figure 2.3.12. Although it was originally designed for unconfined

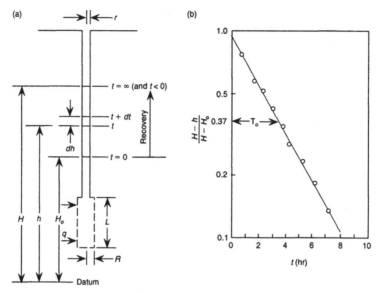

FIG. 2.3.10 Hvorslev piezometer test. (Reprinted from P.B. Bedient et al., 1994, *Groundwater contamination*, PTR Prentice-Hall, Inc.)

aquifers, it can be used for confined aquifers if the top of the screen is some distance below the upper confining layer. The method is based on the following equation:

$$K = \frac{r_c^2 \, \ell n(R_e/r_w)}{2L_e} \frac{1}{t} \ell n \frac{y_o}{y_t} \qquad 2.3(58)$$

where:

R_e = effective radial distance over which the head difference is dissipated

r_w = radial distance between the well center and the undisturbed aquifer (including gravel pack)

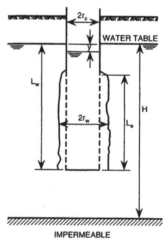

FIG. 2.3.12 Slug test setup. (Reprinted from H. Bouwer, 1978, *Groundwater hydrology*, McGraw-Hill, Inc.)

L_e = height of the perforated, screened, uncased, or otherwise open section of the well through which groundwater enters

y_o = y at time zero

y_t = y at time t

t = time sine y_o

In Equation 2.3(58), y and t are the only variables. Thus, if a number of y and t measurements are taken, they can be plotted on semilog paper to give a straight line. The slope of the best-fitting straight line provides a value for $\ell n(y_o/y_t)/t$. All other parameters in Equation 2.3(58) are known from well geometry, and K can be calculated.

—*Y.S. Chae*

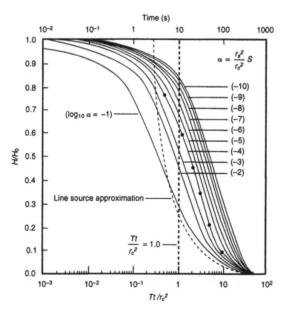

FIG. 2.3.11 Slug test type curves. (Reprinted from I.S. Papadopoulos, J.D. Bredehoeft, and H.H. Cooper, Jr., 1973, On the analysis of slug test data, *Water Resources Res.* 9, no. 4: 1087–1089.)

References

Bedient, P.B. et al. 1994. *Groundwater contamination*, PTR Prentice-Hall, Inc.

Boulton, N.S. 1963. Analysis of data from non-equilibrium pumping tests allowing for delayed yield from storage. *Proc. Inst. Civ. Eng.* 16:469–482.

Bouwer, H. 1978. *Groundwater hydrology.* McGraw-Hill, Inc.

Bouwer, H., and R.C. Rice. 1976. A slug test for determining hydraulic conductivity of unconfined aquifers with completely or partially penetrating wells. *Water Resour. Res.* 12:423–428.

Chow, V.T. 1952. On the determination of transmissivity and storage coefficients from pumping test data. *Trans. Am. Geoph. Union* 33: 397–404.

Cooper, H.H., Jr., J.D. Bredehoeft, and I.S. Papadopoulos. 1967. Response of a finite-diameter well to an instantaneous charge of water. *Water Resour. Res.* 3:263–269.

Cooper, H.H., Jr., and C.E. Jacob. 1946. A generalized graphical method for evaluating formation constants and summarizing well-field history, *Trans. Am, Geoph. Union* 27:526–534.

DeGlee, G.J. 1930. Over grondwaterstromingen bij wateronttrekking door middel van putten. Doctoral dissertation, Techn. Univ., Delft. The Netherlands. Printed by J. Waltman.

DeGlee, G.J. 1951. Berekeningsmethoden voor de winning van grondwater. In *Drinkwaterroorzlening 3e Vacantie cursus*, 38–80. Moorman's periodieke pers. The Hague, Netherlands.

Fetter, C.W. 1988. *Applied hydrogeology.* 2d ed. Macmillan Pub. Co.

Gupta, R.S. 1989. *Hydrology and hydraulic systems.* Prentice-Hall, Inc.

Hantush, M.S. 1956. Analysis of data from pumping tests in leaky aquifers. *Trans. Am. Geophys. Un.* 37:702–714.

Hantush, M.S. 1964. Hydraulics of wells. In *Advances in Hydroscience.* Vol. 1: edited by V.T. Chow, 281–432, New York and London: Academic Press.

Hantush, M.S., and C.E. Jacob. 1955. Non-steady radial flow in an infinite leaky aquifer. *Am. Geophys. Un. Trans.* 36:95–100.

Hvorslev, M.J. 1951. *Time lag and soil permeability in groundwater observations.* U.S. Army Waterways Experiment Station Bull. 36.

Neuman, S.P. 1972. Theory of flow in unconfined aquifers considering delayed response of the water table. *Water Resources Res.* 8:1031–1045.

Neuman, S.P., and P.A. Witherspoon. 1969. Theory of flow in a confine two-aquifer system. *Water Resource Research* 5:803–816.

Papadopoulos, I.S., J.D. Bredehoeft, and H.H. Cooper, Jr. 1973. On the analysis of slug test data. *Water Resources Res.* 9, no. 4:1087–1089.

Walton, W.C. 1962. Selected analytical methods for well and aquifer evaluation. *Illinois State Water Surrey Bull.* 49.

2.4
DESIGN CONSIDERATIONS

In well design, well losses, specific capacity, and partially penetrating wells must be considered.

Well Losses

In the previous sections, the drawdown in a pumping well was assumed to be due only to head losses in the aquifer. In reality, however, additional drawdown is caused by head losses in the well system itself (screen or perforated casing) as water flows through it. The former is known as the drawdown due to the *formation loss* (s_w) and the latter as the drawdown due to the *well loss* (s_f) as shown in Figure 2.4.1. The total drawdown s_t is then $s_t = s_w + s_f$.

Since the flow in the aquifer is laminar, s_w varies linearly with $Q(s_w = C_w Q)$. However, the flow through the well system (screen and perforated casing) is turbulent, and thus s_f can vary with some power of $Q(s_f = C_f Q^n)$. The total drawdown can be expressed as

$$s_t = s_w + s_f = C_w Q + C_f Q^n \qquad 2.4(1)$$

where

$$C_w = \frac{1}{2\pi T} \ell n \left(\frac{R}{r_w} \right) \quad \text{for steady-state} \qquad 2.4(2)$$

$$C_w = \frac{1}{4\pi T} W \left(\frac{r_w^2 S}{4Tt} \right) \quad \text{for unsteady-state} \qquad 2.4(3)$$

$$n = 2 \text{ to } 3.5$$

The best way to determine the values of C_w, C_f, and n is by experiment utilizing a step-drawdown test (Bear 1979;

Bouwer 1978). Well efficiency is related to the well loss and is defined as

$$E_w = \frac{s_w}{s_t} = 1 - \frac{s_f}{s_t} \qquad 2.4(4)$$

Specific Capacity

The *specific capacity* of a well is defined as the well flow per unit drop of water level in the well.

$$\text{Specific capacity} = \frac{Q}{s_t} = \frac{1}{C_w + C_f Q^{n-1}} \qquad 2.4(5)$$

Specific capacity decreases with the pumping rate and time as shown in Figure 2.4.2. It is a useful concept because it describes the productivity of both the aquifer and the well in a single parameter. A reduction of up to 40% in the specific capacity has been observed in one year in wells deriving water entirely from storage (Gupta 1989). The specific capacities of wells in an aquifer system can be used to calculate the transmissivity distribution of the aquifer based on pumping a well of a known diameter for a given period of time.

Partially Penetrating Wells
(Imperfect Wells)

In practice, the underlying impermeable soil layer is often absent or is encountered at a great depth. Wells which do

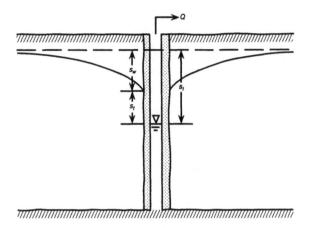

FIG. 2.4.1 Formation and well losses in a pumping well. (Reprinted from R.S. Gupta, 1989, *Hydrology and hydraulic systems*, Prentice-Hall, Inc.)

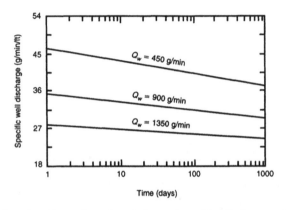

FIG. 2.4.2 Examples of variation of specific discharge with pumping rate and time. (Reprinted from J. Bear, 1979, *Hydraulics of groundwater*, McGraw-Hill, Inc.)

not penetrate through the entire thickness of the aquifer are called *partially penetrating wells* or *imperfect wells*. Imperfect wells are encountered more often in practice than perfect, fully penetrating, wells. Imperfect wells can also have open or closed ends. Figure 2.4.3 shows various configurations of wells in an unconfined aquifer (ordinary

wells), and Figure 2.4.4 shows configurations for wells in a confined aquifer (artesian wells).

Compared to a fully penetrating well, a partially penetrating well has an additional head loss, as shown in Figure 2.4.5, due to the convergence of flow lines and their extended length. Hence, for a given pumping rate Q, the drawdown in an imperfect well is larger than in a perfect well.

Starting with Forchheimer in 1898, numerous analytical and empirical equations have been developed to solve imperfect wells: Kozney (1933), Muskat (1937), Li et al, Hantush (1962), and Kirkham (1959) among others.

CONFINED AQUIFERS

As previously stated, an imperfect well requires more drawdown for a given Q than does a perfect well. The additional drawdown can be represented by

$$s_{wp} = \frac{Q}{4\pi T}\left(\ell n \frac{2.25Tt}{r^2 S} + 2s_p\right) \qquad 2.4(6)$$

where s_{wp} is the drawdown at the well. The values of s_p as a function of H/r_w for various values of L_e/H can be obtained from Figure 2.4.6, which was developed by Sternberg (1967).

The performance of an imperfect well related to a perfect well is expressed as an efficiency, defined as the ratio of Q_p to Q at a given drawdown as

$$\frac{Q_p}{Q} = \frac{1}{1 + \dfrac{s_p}{s_w}\dfrac{Q}{2\pi T}} \qquad 2.4(7)$$

or

$$\frac{Q_p}{Q} = \frac{1}{1 + \dfrac{s_p}{\ell n\left(\dfrac{r}{r_w}\right)}} \qquad 2.4(8)$$

Equation 2.4(8) applies to wells with their perforated or open section at the top (Figure 2.4.5-c) or at the bot-

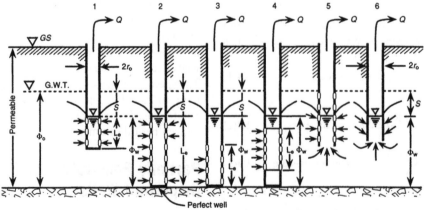

FIG. 2.4.3 Types of ordinary imperfect wells. (Reprinted from A.R. Jumikis, 1964, *Mechanics of soils*, Van Nostrand.)

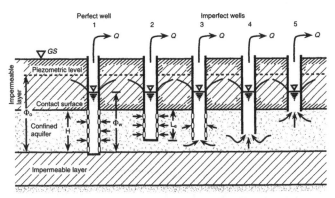

FIG. 2.4.4 Types of artesian imperfect wells. (Reprinted from A.R. Jumikis, 1964, *Mechanics of soils*, Van Nostrand.)

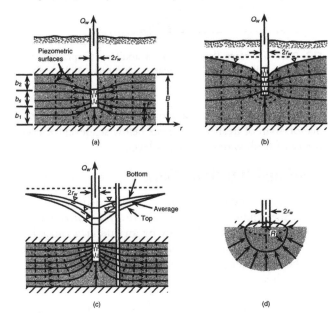

FIG. 2.4.5 Partially penetrating wells. (a) In a confined aquifer. (b) In a phreatic aquifer. (c) Drawdown curves along streamlines. (d) Zero penetration in a thick aquifer. (Reprinted from J. Bear, 1979, *Hydraulics of groundwater*, McGraw-Hill, Inc.)

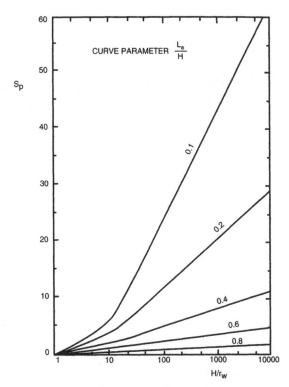

FIG. 2.4.6 Graph of s_p versus H/r_w. (Reprinted from Y.M. Sternberg, 1967, Efficiency of partially penetrating wells, *Ground Water* II, no. 3:5–8.)

$$\frac{Q_p}{Q} = \sqrt{\frac{L_e}{z}}\, 4\sqrt{\frac{2z - L_e}{z}} \quad \text{for closed end} \qquad 2.4(9)$$

and

$$\frac{Q_p}{Q} = \sqrt{\frac{L_e + 0.5r_w}{z}} \cdot 4\sqrt{\frac{2z - L_e}{z}} \quad \text{for open end} \qquad 2.4(10)$$

where z = distance from the water level in the well to the impervious stratum in meters.

—Y.S. Chae

References

Bear, J. 1979. *Hydraulics of groundwater*. McGraw-Hill, Inc.

Bouwer, H. 1978. *Groundwater hydrology*. McGraw-Hill, Inc.

Forchheimer, Ph., 1898. Grundwasserspiegel bei Brunnenanlagen. *Zeitschrift des Osterreichischen Ingenieur-und Architekten-Vereins* 50, no. 45:645.

Gupta, R.S. 1989. *Hydrology and hydraulic systems*. Prentice-Hall, Inc.

Hantush, M.S. 1962. Aquifer tests on partially penetrating wells. Trans. Am. Soc. Civ. Eng. Vol. 127, pt. 1:284–308.

Kirkham, D. 1959. Exact theory of flow into a partially penetrating well. *J. Geophys. Research* 64:1317–1327.

Kozeny, J. 1933. Theorie und Berechnung der Brunnen. *Wasserkraft und Wasserwirtschaft* 28:104.

Li, W.H. et al. A new formula for flow into partially penetrating wells in aquifers. Trans. Am. Geophys. Union 35:806–811.

Muskat, M. 1937. *The flow of homogeneous fluids through porous media*. McGraw-Hill.

Sternberg, Y.M. 1967. Efficiency of partially penetrating wells. *Ground Water* 11, no. 3:5–8.

tom. If the open section of the well is in the center of the aquifer (Figure 2.4.5-a), vertical flow components occur at both the top and bottom of the section. For this case, Q_p/Q can be obtained for the half-section when the section is split symmetrically along the midway.

UNCONFINED AQUIFERS

The s_p values in Figure 2.4.6 also give reasonable estimates of Q_p/Q for wells in unconfined aquifers, particularly if the drawdown is small compared to the aquifer thickness and the well has been pumped for some time (Bouwer 1978).

Forchheimer (1898) observed that the discharge Q_p of a well with a perforated casing and closed end becomes larger, at equal drawdowns, as the immersed depth L_e of the perforated well increases. He assumed that the increase with depth varies with the geometric mean between the parabolic and elliptic ordinates and showed an efficiency to be

2.5
INTERFACE FLOW

The flow of groundwater in coastal aquifers, as shown in Figure 2.5.1, can be treated as an interface flow problem in which two fluids of different densities, fresh and salt water, have a clear interface rather than a transition zone. This flow problem assumes that the fresh water flows over the salt water which is at rest. These flows are denoted as the Ghyben–Duipint approximations.

The pressure distribution in the salt water ρ_s is

$$p_s = \rho_s g(\phi_s - z) \qquad 2.5(1)$$

and the pressure distribution in the fresh water p_f is

$$p_f = \rho_f g(\phi - z) \qquad 2.5(2)$$

where ϕ_s and ϕ are the head in the salt and fresh water respectively, and z is the distance from the reference plane to the interface. The pressure at any point of the interface must be a single value, that is $p_f = p_s$. Therefore, with Equations 2.5(1) and 2.5(2) and with $z = H_s - h_s$, then

$$\rho_s g(\phi_s - H_s + h_s) = \rho_f g(\phi - H_s + h_s) \qquad 2.5(3)$$

If $\phi_s = H_s$ and $\phi = H_s + h_f$, Equation 2.5(3) yields

$$h_s = \frac{\rho_f}{\rho_s - \rho_f} h_f = \alpha h_f \qquad 2.5(4)$$

This equation is known as the Ghyben–Herzberg equation. This equation is also valid for confined aquifers, in which the upper boundary of the aquifer is a horizontal impermeable boundary rather than a phreatic surface and h_f represents the piezometric head with respect to sea level. The ratio between the densities of salt and fresh water is of the order of 1.025. Then, Equation 2.5(4) shows that

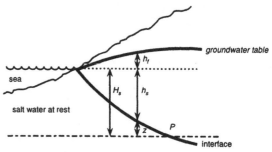

FIG. 2.5.1 Interface flow in coastal aquifers. (Reprinted from O.D.L. Strack, 1989, *Groundwater mechanics,* Vol. 3, Pt. 3, Prentice-Hall, Inc.)

h_s is about 40 times h. Therefore, in coastal aquifers, storage of 40 m of fresh water exists below sea level for every meter of fresh water above sea level.

Confined Interface Flow

Figure 2.5.2 shows the shallow confined interface flow when the aquifer is bounded above by a horizontal impervious boundary and below by an interface. Since $h = h_s - (H_s - H)$ and the head $\phi = h_f + H_s$, use of the Ghyben–Herzberg equation gives the thickness of the aquifer h as

$$h = \frac{\rho_f}{\rho_s - \rho_f}\phi - \frac{\rho_s}{\rho_s - \rho_f}H_s + H \qquad 2.5(5)$$

The elevation z of the interface above the reference level equals $z = H_s - h_s$. Use of the Ghyben–Herzberg equation then yields

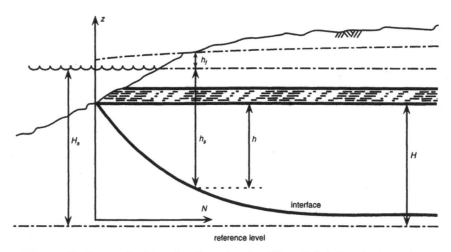

FIG. 2.5.2 Shallow confined interface flow. (Reprinted from O.D.L. Strack, *Groundwater mechanics,* Vol. 3, Pt. 3, Prentice-Hall, Inc.)

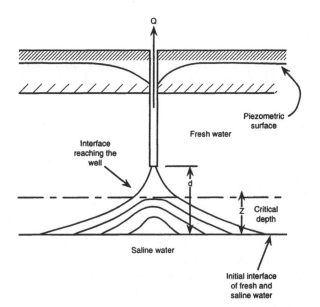

FIG. 2.5.3 Upconing of salt water under a pumping well. (Reprinted from R.S. Gupta, 1989, *Hydrology and hydraulic systems,* Prentice-Hall, Inc.)

$$z = \frac{\rho_s}{\rho_s - \rho_f} H_s - \frac{\rho_f}{\rho_s - \rho_f} \phi \qquad 2.5(6)$$

Unconfined Interface Flow

A shallow unconfined aquifer is shown in Figure 2.5.2. The aquifer thickness h can now be expressed with the Ghyben–Herzberg equation as

$$h = \frac{\rho_s}{\rho_s - \rho_f} \phi - \frac{\rho_s}{\rho_s - \rho_f} H_s \qquad 2.5(7)$$

and the elevation of the interface z is obtained in the same way as the confined interface flow as

$$z = \frac{\rho_s}{\rho_s - \rho_f} H_s - \frac{\rho_f}{\rho_s - \rho_f} \phi \qquad 2.5(8)$$

Upconing of Saline Water

Figure 2.5.3 depicts a situation in which water is pumped from a freshwater zone underlaid by a saline water layer. The interface between fresh water and saline water rises toward the well in a cone shape as shown in the figure. This phenomenon is known as *upconing*.

The height of upconing under the steady-state condition (Gupta 1989) is given by

$$z = \frac{Q}{2\pi Kd} \frac{\rho_f}{\rho_s - \rho_f} \qquad 2.5(9)$$

where d = depth to the initial interface below the bottom

TABLE 2.5.1   CONTROLLING SALTWATER INTRUSION OF VARIOUS CATEGORIES

Source or Cause of Intrusion	Control Methods
Seawater in coastal aquifer	Modification of pumping pattern
	Artificial recharge
	Extraction barrier
	Injection barrier
	Subsurface barrier
Upconing	Modification of pumping pattern
	Saline scavenger wells
Oil field brine	Elimination of surface disposal
	Injection wells
	Plugging of abandoned wells
Defective well casings	Plugging of faulty wells
Surface infiltration	Elimination of source
Saline water zones in freshwater aquifers	Relocation and redesign of wells

Source: D.K. Todd, 1980, *Groundwater hydrology* (John Wiley and Sons).

of the well. Salt water reaches the well, contaminating the supply, when the rise becomes critical at z = 0.3 to 0.5 d. Thus, the maximum discharge that keeps the rise below the critical limit is obtained when z = 0.5 d is substituted in Equation 2.5(9) as

$$Q_{max} = \pi K d^2 \frac{\rho_s - \rho_f}{\rho_f} \qquad 2.5(10)$$

In reality, brackish water occurs between fresh and salt water. Even with a low rate of pumping, some saline water inevitably reaches the pump. Increasing the distance d and decreasing the rate of pumping Q minimizes the upconing effect.

Protection Against Intrusion

Controlling the intrusion of saline water before it contaminates an aquifer system is desirable because removing it once it has developed is difficult. Years may be required to restore normal conditions. Table 2.5.1 summarizes many control methods suggested for various categories of problems.

—*Y.S. Chae*

Reference

Gupta, R.S. 1989. *Hydrology and hydraulic systems.* Prentice-Hall, Inc.

3

Principles of Groundwater Contamination

3.1
CAUSES AND SOURCES OF CONTAMINATION

A groundwater contaminant is defined by most regulatory agencies as any physical, chemical, biological, or radiological substance or matter in groundwater. The contaminants can be introduced in the groundwater by naturally occurring activities, such as natural leaching of the soil and mixing with other groundwater sources having different chemistry. They are also introduced by planned human activities, such as waste disposal, mining, and agricultural operations. Because the contamination from naturally occurring activities is usually small, it is not the focus of this chapter. However, human activities are the leading cause of groundwater contamination and the focus of most regulatory agencies.

The most prevalent human activities that cause groundwater contamination are (1) waste disposal, (2) storage and transportation of commercial materials, (3) mining operations, (4) agricultural operations, and (5) other activities as shown in Figure 3.1.1.

This section discusses the principal sources and causes of groundwater contamination from these activities with regard to their occurrence and effects on groundwater quality.

Waste Disposal

Waste disposal includes the disposal of liquid waste and solid waste.

LIQUID WASTE

Underground or aboveground disposal practices of domestic, municipal, or industrial liquid waste can cause groundwater contamination. Among all disposal practices of domestic liquid waste, septic tanks and cesspools contribute the most wastewater to the ground and are the most frequently reported sources of groundwater contamination (U.S. EPA 1977). Septic tanks and cesspools contribute filtered sewage effluent directly to the ground which can introduce high concentrations of BOD, COD, nitrate, organic chemicals, and possibly bacteria and viruses into groundwater (Mallmann and Mack 1961; Miller 1980). Also, chlorination of the wastewater effluent and the use of chemicals to clean septic systems can produce additional potential pollutants (Council on Environmental Quality 1980).

With regard to municipal liquid waste, land application of sewage effluent and sludge is perhaps the largest contributor to groundwater contamination. Treated wastewater and sludge have been applied to land for many years to recharge groundwater and provide nutrients that fertilize the land and stimulate plant growth (Bauer 1974; U.S. EPA 1983). However, land application of sewage effluent can introduce bacteria, viruses, and organic and inorganic chemicals into groundwater (U.S. EPA 1974).

Another major municipal source of groundwater contamination is urban runoff from roadway deicing. In many urban areas, large quantities of salts and deicing additives are applied to roads during the winter months. These salts and additives facilitate the melting of ice and snow; however, they can percolate with the water into the ground and cause groundwater contamination of shallow aquifers (Field et al. 1973). In addition, the high solubility of these salts in water and the relatively high mobility of the resulting contaminants such as chloride ions in groundwater can cause the zone of contamination to expand (Terry 1974).

With regard to industrial liquid waste, surface impoundments and injection wells are probably the largest contributors to groundwater contamination. As legislation to protect surface water resources has become more stringent, the use of surface impoundments and injection wells has become an attractive wastewater effluent disposal option for many industries. However, leakage of contaminants through the bottom of a surface impoundment or migration of fluids from an injection well into a hydrologically connected usable aquifer can cause groundwater contamination (Council on Environmental Quality 1981). The extent and severity of groundwater contamination from these sources is further complicated by the fact that, in addition to being hazardous, many of the organic and

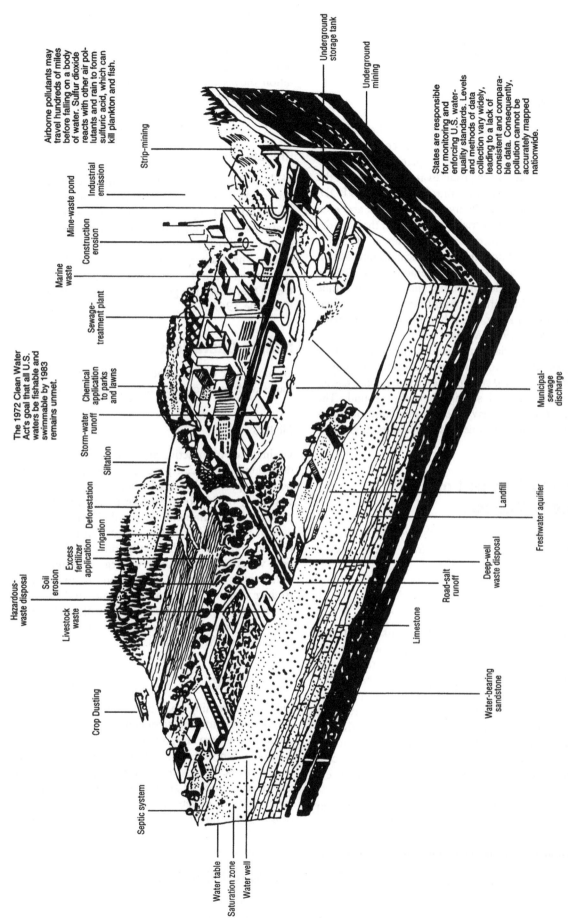

FIG. 3.1.1 Sources of groundwater contamination. (Reprinted from National Geographic, 1993, Water, *National Geographic Special Editions* [November], Washington, D.C.: National Geographic Society.)

Airborne pollutants may travel hundreds of miles before falling on a body of water. Sulfur dioxide reacts with other air pollutants and rain to form sulfuric acid, which can kill plankton and fish.

States are responsible for monitoring and enforcing U.S. water-quality standards. Levels and methods of data collection vary widely, leading to a lack of consistent and comparable data. Consequently, pollution cannot be accurately mapped nationwide.

The 1972 Clean Water Act's goal that all U.S. waters be fishable and swimmable by 1983 remains unmet.

Underground storage tank

Underground mining

Strip-mining

Industrial emission

Mine-waste pond

Construction erosion

Marine waste

Sewage-treatment plant

Chemical application to parks and lawns

Storm-water runoff

Siltation

Deforestation

Irrigation

Excess fertilizer application

Soil erosion

Hazardous-waste disposal

Livestock waste

Crop Dusting

Septic system

Water table

Saturation zone

Water well

Municipal-sewage discharge

Landfill

Freshwater aquifer

Deep-well waste disposal

Road-salt runoff

Limestone

Water-bearing sandstone

inorganic chemicals in industrial wastewater effluent and sludge are persistent.

SOLID WASTE

The land disposal of municipal and industrial solid waste is another potential cause of groundwater contamination. Buried waste is subject to leaching by percolating rain water and surface water or by groundwater contact with the fill. The generated leachate can contain high levels of BOD, COD, nitrate, chloride, alkalinity, trace elements, and even toxic constituents (in industrial waste landfill) that can degrade the quality of groundwater (Hughes et al. 1971; Zanoni 1972). In addition, the biochemical decomposition of the organic matter in waste generates gases such as methane, carbon dioxide, ammonia, and hydrogen sulfide that can migrate through the unsaturated zone into adjacent terrains and cause potential hazards such as methane explosions (Flower 1976; Mohsen 1975).

Stockpiles of materials and waste tailings can also be a source of groundwater contamination. Precipitation falling on uncovered or unlined stockpiles or waste tailings causes leachate generation and seepage into the ground. The leachate can transport heavy metals, salts, and other inorganic and organic constituents as pollutants to groundwater.

Storage and Transport of Commercial Materials

Groundwater contamination from the storage and transport of commercial materials results from leaking storage tanks and spills.

STORAGE TANKS

Underground and aboveground storage tanks and transmission pipelines are another cause of groundwater pollution. Among all underground storage tanks and pipelines, gasoline and home oil fuel tanks probably contribute the most to groundwater contamination. These tanks and pipelines are subject to corrosion and structural failures with subsequent leaks that introduce a variety of contaminants into groundwater. Leakage is particularly frequent from bare steel tanks that are not protected against corrosion. Even if a leakage is small, it can pose a significant threat to groundwater quality.

Gasoline and petroleum products contain hydrocarbon components such as benzene, toluene, and xylene that are highly soluble and mobile in groundwater and can be hazardous to humans if consumed. One gallon of gasoline is enough to render one million gallons of groundwater unusable based on U.S. Environmental Protection Agency (EPA) drinking water standards (Noonan and Curtis 1990). In addition, vapors and immiscible compounds

trapped in the pore spaces of the unsaturated zone continue to feed groundwater with contaminants as precipitation moves into and through the subsurface or as the groundwater table fluctuates (Dietz 1971; Van Dam 1967).

SPILLS

Spills and discharges on the ground of chemical products can migrate downward and contaminate groundwater. Spills and discharges vary from casual activities at industrial sites, such as leaks from pipes and valves, to accidents involving aboveground storage tanks, railroad cars, and trucks. The discharged chemicals are usually entrained by stormwater runoff and transported to the subsurface where they reach the groundwater and degrade its quality (Scheville 1967).

Mining Operations

Groundwater can be contaminated by the drainage from mines and by oil and gas mining operations.

MINES

Drainage of both active and abandoned surface and underground mines can produce a variety of groundwater pollution problems (Emrich 1969). Rainwater, particularly acid rain, overexposed surface mines, and mine tailings produce highly mineralized runoff frequently referred to as *acid mine drainage*. This runoff can percolate into the ground and degrade the quality of groundwater. In addition, water seepage through underground mines can leach toxic metals from exposed ores and raw materials and introduce them to groundwater (Barnes and Clarke 1964). Oxidation and leaching connected with coal mining produce high iron and sulfate concentrations and low pH in groundwater (Miller 1980).

OIL AND GAS

Oil and gas mining operations can also cause groundwater contamination. These operations generate a substantial amount of wastewater, often referred to as *brine*. The brine is usually disposed of in surface impoundments or injected in deep wells. Therefore, it can reach groundwater, and its constituents, such as ammonia, boron, calcium, dissolved solids, sodium, sulfate, and trace metals, can subsequently degrade the quality of groundwater (Fryberger 1975; Warner 1965).

Agricultural Operations

The use of fertilizers and pesticides in agricultural operations can contaminate groundwater.

FERTILIZERS

Fertilizers are the primary cause of groundwater contamination beneath agricultural lands. Both inorganic (chemically manufactured) and organic (from animal or human waste) fertilizers applied to agricultural lands provide nutrients such as nitrogen, phosphorous, and potassium that fertilize the land and stimulate plant growth. A portion of these nutrients usually leaches through the soil and reaches the groundwater table. Phosphate and potassium fertilizers are readily adsorbed on soil particles and seldom constitute a pollution problem. However, only a portion of nitrogen is adsorbed by soil or used by plants, and the rest is dissolved in water to form nitrates in a process called *nitrification*. Nitrates are mobile in groundwater and have potential to harm infant human beings and livestock if consumed on a regular basis (Hassan 1974).

PESTICIDES

Pesticides, herbicides, and fungicides used for destroying unwanted animal pests, plants, and fungal growth can also cause groundwater contamination. When applied to land or disposed of in landfills, these chemicals degrade in the environment by a variety of mechanisms. However, their parent compounds and their byproducts persist long enough to adversely impact the soil and groundwater (California Department of Water Resources 1968).

Other Activities

Interaquifer exchange and saltwater intrusion are two other human activities that cause groundwater contamination.

INTERAQUIFER EXCHANGE

In interaquifer exchange, two aquifers are hydraulically connected. Contamination occurs when contaminants are transferred from a contaminated aquifer to a clean aquifer. Interaquifer exchange is common when a deep well penetrates more than one aquifer to provide increased yield or when an improperly cased or abandoned well serves as a direct connection between two aquifers of different potential heads and different water quality. The hydraulic connection (well or fractures) can allow contaminants from aquifers with the greatest hydraulic head to move to aquifers of less hydraulic head (Deutsch 1961).

SALTWATER INTRUSION

Saltwater intrusion, in which saline water displaces or mixes with fresh groundwater, is another source of groundwater contamination. Saltwater intrusion is usually caused when the hydrodynamic balance between the fresh water and the saline water is disturbed, such as when fresh groundwater is overpumped in coastal aquifers (Task Committee on Salt Water Intrusion 1969). Saltwater intrusion can also occur when the natural barriers that separate fresh and saline water are destroyed, such as in the construction of coastal drainage canals that enable tidal water to advance inland and percolate into a freshwater aquifer (Todd 1974).

—*Ahmed Hamidi*

References

Barnes, I., and F.E. Clarke. 1964. *Geochemistry of groundwater in mine drainage problems.* U.S. Geotechnical Survey Prof. Paper 473-A.

Bauer, W.J. 1974. Land treatment designs, present and future. Proceedings of the International Conference on Land for Waste Management, edited by J. Thomlinson. 343–346. Ottawa, Canada: National Research Council.

California Department Water Resources. 1968. *The fate of pesticides applied to irrigated agricultural land* Bv11.174-1. Sacramento, Calif.

Council on Environmental Quality. 1980. *The eleventh annual report of the Council on Environmental Quality.* December.

———. 1981. Contamination of groundwater by toxic organic chemicals. (January). Washington, D.C.: U.S. Government Printing Office.

Deutsch, M. 1961. Incidents of chromium contamination of groundwater in Michigan. Proceedings of Symposium on Groundwater Contamination, April. Cincinnati, Ohio: U.S. Dept. of Health, Education and Welfare.

Dietz, D.N. 1971. Pollution of permeable strata by oil components. In *Water pollution by oil,* edited by Peter Hepple, 128–142. Elsevier, Amsterdam.

Emrich, G.H., and G.L. Merritt. 1969. Effects of mine drainage on groundwater. *Groundwater* 7, no. 3:27–32.

Field, R. et al. 1973. *Water pollution and associated effects from street salting.* EPA-R2-73-257. Cincinnati, Ohio: U.S. EPA.

Flower, F.B. 1976. *Case history of landfill gas movement through soils,* edited by E.J. Genetilli and J. Cirello, 177–184. Cincinnati, Ohio: U.S. EPA.

Fryberger, J.S. 1975. Investigation and rehabilitation of a brine-contaminated aquifer. *Groundwater* 13, no. 2:155–160.

Hassan, A.A. 1974. Water quality cycle—reflection of activities of nature and man. *Groundwater* 12, no. 1:16–21.

Hughes, G. et al. 1971. *Pollution of groundwater due to municipal dumps.* Tech. Bull. no. 42. Ottawa, Ont.: Canada Dept. of Energy, Mines and Resources, Inland Waters Branch.

Mallmann, W.L., and W.N. Mack. 1961. Biological contamination of groundwater. Proceedings of Symposium on Groundwater Contamination. April. U.S. Department of Health, Education and Welfare.

Miller, D.W. 1980. *Waste disposal effects on groundwater.* Berkeley, Calif.: Premier Press.

Mohsen, M.F.N. 1975. Gas migration from sanitary landfills and associated problems. Ph.D. thesis, University of Waterloo.

Noonan, D.C., and J.T. Curtis. 1990. *Groundwater remediation and petroleum: A guide for underground storage tanks.* Chelsea, Mich.: Lewis Publishers.

Scheville, F. 1967. Petroleum contamination of the subsoil, a hydrological problem. In *The joint problems of the oil and water industries,* edited by Peter Hepple. 23–53. Elsevier, Amsterdam.

Task Committee on Salt Water Intrusion. 1969. Saltwater intrusion in the United States. *Journal of Hydraulics Division, ASCE* 95, no. Hy5:1651–1669.

Terry, R.C., Jr. 1974. *Road salt, drinking water, and safety.* Cambridge, Mass.: Ballinger.

Todd, D.K. 1974. Salt water intrusion and its controls. *Journal of AWWA* 66:180–187.

U.S. Environmental Protection Agency. 1974. Land application of sewage effluents and sludge, selected abstracts. Washington, D.C.: Government Printing Office.

———. 1977. *Waste disposal practices and their effects on groundwater.* Report to Congress, 81–107.

———. 1983. Process design manual: Land application of municipal

sludge. EPA-625/1-83-016. Cincinnati, Ohio: U.S. EPA, Municipal Environmental Lab.

Van Dam, J. 1967. The migration of hydrocarbons in water bearing stratum. In *The joint problems of the oil and water industries,* edited by Peter Hepple. London: Institute of Petroleum.

Warner, D.L. 1965. *Deep-well injection of liquid waste.* Publ. no. 999-WP-21. U.S. Public Health Service.

Zanoni, A.E. 1972. Groundwater pollution and sanitary landfills—a critical review. *Groundwater* 10, no. 1:3–16.

3.2
FATE OF CONTAMINANTS IN GROUNDWATER

When a contaminant is introduced in the subsurface environment, its fate and concentration are controlled by a variety of physical, chemical, and biochemical processes that occur between the contaminant and the constituents of the subsurface environment. A complete discussion and assessment of all these processes for all contaminants are beyond the scope of this chapter. However, this section illustrates some of the most important processes for several groups of contaminants and the impact of these processes on the concentration and mobility of contaminants.

Organic Contaminants

The physicochemical reactions that can alter the concentration of an organic contaminant in groundwater can be grouped into five categories as suggested by Arthur D. Little (1976) and Rao and Jessup (1982). These categories include (1) hydrolysis of the contaminant in water, (2) oxidation–reduction, (3) biodegradation of the contaminant by microorganisms, (4) adsorption of the contaminant by the soil, and (5) volatilization of the contaminant to the air present in the unsaturated zone. The relative importance of each of these reactions depends on the physical and chemical characteristics of the contaminant and on the specific conditions of the subsurface environment.

HYDROLYSIS

Hydrolysis is a chemical reaction in which an organic chemical (RX) reacts with water or a hydroxide ion (OH) as follows:

$$R - X + H_2O \longrightarrow R - OH + H^+ + X^- \qquad 3.2(1)$$

$$R - X + OH^- \longrightarrow R - OH + X^- \qquad 3.2(2)$$

During these reactions, a leaving group (X) is replaced by a hydroxyl ion (OH), and a new carbon–oxygen bond is

formed. The R represents the carbonium ion and the X the leaving group. Common leaving groups include halides (Cl^-, Br^-), alcohols ($R—O^-$), and amines ($R_1R_2N^-$). The acquisition of a new polar functional group increases the water solubility of the organic chemical.

Examples of hydrolysis include the following (Valentine 1986):

$$\begin{array}{lll} RCl + H_2O & \longrightarrow ROH + H^+ + Cl^- & 3.2(3) \\ \text{an alkyl halide} & \text{an alcohol} & \end{array}$$

$$\begin{array}{lll} R_1COOR_2 + H_2O & \longrightarrow R_2OH + R_1COOH & 3.2(4) \\ \text{an ester} & \text{an alcohol + a carboxylic acid} & \end{array}$$

$$\begin{array}{lll} RC(ON)R_1R_2 + H_2O \longrightarrow RCOOH + R_1R_2NH & 3.2(5) \\ \text{an amide} & \text{a carboxylic acid + an amine} & \end{array}$$

$$\begin{array}{lll} RCH_2CN + H_2O & \longrightarrow RCH_2COOH + NH_3 & 3.2(6) \\ \text{a nitrile} & \text{a carboxylic acid + ammonia} & \end{array}$$

The hydrolysis of organic chemicals in water is generally considered first-order with respect to the organic chemical's concentration; thus, the rate of hydrolysis can be calculated with the following equation (Dragun 1988b):

$$k \cdot C = -\frac{dC}{dt} \qquad 3.2(7)$$

or

$$k = \frac{2.303}{t} \log\left(\frac{C_0}{C}\right) \qquad 3.2(8)$$

where:

k = rate constant, 1/time
t = time
C_0 = initial concentration, ppm
C = concentration at time t, ppm

The time needed for half of the concentration to react, half-life, can be calculated if k is known with use of the following equation:

$$t_{1/2} = \frac{0.693}{k} \qquad 3.2(9)$$

where $t_{1/2}$ is equal to the half-life.

Table 3.2.1 lists the hydrolysis half-lives for several organic chemicals. Half-lives vary from seconds to tens of thousands of years. Certain compounds such as alkyl halides, chlorinated amides, amines, carbamates, esters, epoxides, phosphonic acid esters, phosphoric acid esters, and sulfones are potentially susceptible to hydrolysis (Dragun 1988b). Other compounds such as aldehydes, alkanes, alkenes, alkynes, aliphatic amides, aromatic hydrocarbons and amines, carboxylic acids, and nitro fragments are generally resistant to hydrolysis (Dragun 1988b; Harris 1982).

For organic chemicals undergoing an acid- and base-catalyzed hydrolysis (in the case of acid or alkaline solutions), the total hydrolysis rate constant k_T can be expressed (Harris 1982; Mabey and Mill 1978) as

$$k_T = k_H [H^+] + k_N + k_{OH} [OH^-] \qquad 3.2(10)$$

where:

k_T	= total hydrolysis rate constant
k_H	= rate constant for acid-catalyzed hydrolysis
$[H^+]$	= hydrogen ion concentration
$[OH^-]$	= hydroxyl ion concentration
k_N	= rate constant for neutral hydrolysis
k_{OH}	= rate constant for base-catalyzed hydrolysis

Several other parameters can affect the rate of hydrolysis including temperature, the pH of the soil particle surfaces, the presence of metals in soils, the adsorption of the organic chemical, and the soil water content (Burkhard and Guth 1981; Konrad and Chesters 1969).

After the hydrolysis rate constant k is estimated, the behavior of a compound can be modeled with a form of the advection–dispersion equation. Equation 3.3(1), that includes a first-order degradation term.

OXIDATION–REDUCTION

In organic chemistry, oxidation–reduction (redox) refers to the transfer of atoms rather than direct electrons as is the case of inorganic chemistry. Oxidation of an organic compound frequently involves a gain in oxygen and a loss in hydrogen atoms, and the reduction involves a gain in hydrogen and a loss in oxygen content. Oxidation–reduction reactions greatly affect contaminant transport and are usually closely related to the microbial activity and the type of substrates available to the organisms. Organic contaminants provide the reducing equivalents for the microbes. After the oxygen in the subsurface environment is depleted, the most easily reduced materials begin to react and, along with the reduced product, dictate the dominant potential.

The occurrence of oxidation in the subsurface is a function of the electrical potential in the reacting system (Dragun and Helling 1985). For oxidation to occur, the potential of the soil system must be greater than that of the organic chemical. Soil reduction potentials can be generally classified as follows:

Aerated soils:	+0.8 to +0.4 volts
Moderately reduced soils:	+0.4 to +0.1 volts
Reduced soils:	+0.1 to −0.1 volts
Highly reduced soils:	−0.1 to −0.5 volts

A number of organic chemicals can hydrolyze, oxidize, and reduce quickly and sometimes violently upon contact with groundwater. Table 3.2.2 lists several classes of organic chemicals that react rapidly and violently with groundwater.

The hydrolysis, oxidation, or reduction of one organic chemical usually results in the synthesis of one or more new organic chemicals. Organic chemistry textbooks identify the basic reaction products. Soil minerals can significantly influence the chemical structure of reaction products. In addition, certain organic chemicals can form significant amounts of residues that bind to the soil. Examples of such chemicals include anilines, phenols, triazines, urea herbicides, carbamates, organophosphates, and cyclodiene insecticides (Sax 1984).

BIODEGRADATION

Biodegradation of an organic chemical in soil is the modification or decomposition of the chemical by soil microorganisms to produce ultimately microbial cells, carbon dioxide, oxygen, and water.

Soil serves as the home for numerous microorganisms capable of degrading organic chemicals. The most predominant microorganisms in soil include bacteria, actinomycetes, and fungi. One gram of surface soil can contain from 0.1 to 1 billion cells of bacteria, 10 to 100 million cells of actinomycetes, and 0.1 to 1 million cells of fungi (Dragun 1988b). The microorganism population in soils is generally greatest in the surface horizons where temperature, moisture, and energy supply is favorable for their growth. As the depth increases, the number of aerobic microorganisms decreases; however, anaerobic microorganisms can exist depending on availability of nutrients and organic material.

The biodegradation of an organic chemical by a microorganism is catalyzed by enzymes which are produced as part of the metabolic activity of the living organism. The biotransformation occurs either inside the microorganism via intracellular enzymes or outside the microorganism by the action of extracellular enzymes. After an organic chemical and an enzyme collide, an enzyme–chemical complex forms. Then, depending on the alignment between the functional groups of the chemical and the enzyme, a reaction product (modified or decomposed organic chemical) is formed by the removal of one or more functional groups by oxidation or reduction reactions (Dragun

TABLE 3.2.1 HYDROLYSIS HALF-LIVES FOR VARIOUS ORGANIC CHEMICALS

Chemical	Half-life	Chemical	Half-life
acetamide	3,950 y	ethion	9.9 d
atrazine	2.5 h	N-ethylacetamide	70,000 y
azirdine	154 d	ethyl acetate	136 d
benzoyl chloride	16 s	ethyl butanoate	5.8 y
benzyl bromide	1.32 h	ethyl trans-buteonate	17 y
benzyl chloride	15 h	ethyl difluoroethanoate	23 m
benzylidene chloride	0.1 h	ethyl dimethylethanoate	9.6 y
bromoacetamide	21,200 y	ethyl methylthioethanoate	87 d
bromochloromethane	44 y	ethyl phenylmethanoate	7.3 y
bromodichloromethane	137 y	ethyl propanoate	2.5 y
bromoethane	30 d	ethyl propenoate	3.5 y
1-bromohexane	40 d	ethyl propynoate	17 d
3-bromohexane	12 d	ethyl pyridylmethanoate	0.41 y
bromomethane	20 d	fluoromethane	30 y
bromomethylepoxyethane	16 d	2-fluor-2-methylpropane	50 d
1-bromo-3-phenylpropane	290 d	hydroxymethylpropane	28 d
1-bromopropane	26 d	iodoethane	49 d
3-bromopropene	12 h	iodomethane	110 d
chloroacetamide	1.46 y	2-iodopropane	2.9 d
chlorodibromoethane	274 y	3-iodopropene	2.0 d
chloroethane	38 d	isobutyramide	7,700 y
chlorofluoriodomethane	1.0 y	isopropyl bromide	2.0 d
chloromethane	339 d	isopropyl ethanoate	8.4 y
chloromethylepoxyethane	8.2 d	malathion	8.1 d
2-chloro-2methylpropane	23 s	methoxyacetanide	500 y
2-chloropropene	2.9 d	N-methylacetamide	38,000 y
3-chloropropene	69 d	methyl chloroethanoate	14 h
cyclopentanecarboxamide	5,500 y	methyl dichloroethanoate	38 m
dibromoethane	183 y	methylepoxyethane	14.6 d
1,3-dibromopropane	48 d	methyl parathion	10.9 d
dichloroacetamide	0.73 y	methyl trichloroethanoate	<3.6 m
dichloroiodomethane	275 y	monomethyl phosphate	1.0 d
dichloromethane	704 y	parathion	17 d
dichloromethyl ether	25 s	phenyl dichloroethanoate	3.7 m
diethyl methylphosphonate	990 y	phenyl ethanoate	38 d
dimethoxysulfone	1.2 m	phosphonitrilic hexamide	46 d
1,2-dimethylepoxyethane	15.7 d	propadienyl ethanoate	110 d
1,1-dimethylepoxyethane	4.4 d	ronnel	1.6 d
diphenyl phosphate	20.6 d	tetrachloromethane	7,000 y (1 ppm)
epoxyethane	12 d	tribromomethane	686 y
3,4 epoxycyclohexene	6 m	trichloroacetamide	0.23 y
3,4 epoxycyclooctane	52 m	trichloromethane	3,500 y
1,3-epoxy-1-oxopropane	3.5	trichlorethylbenzene	19 s
		triethyphosphate	5.5 y
		tri(ethylthio)phosphate	8.5 y

Source: J. Dragun, 1988, *The soil chemistry of hazardous materials* (Silver Springs, MD: Hazardous Materials Control Research Institute).
d = days, h = hours, s = seconds, m = minutes, y = years.

1988a). Some typical biochemical reactions are as follows (Valentine and Schnoor 1986):

Decarboxylation:

$$R—OCH_3 \longrightarrow ROH + CO_2 \qquad 3.2(11)$$

Oxidation of an amino group:

$$R—NH_2 \longrightarrow RNO_2 \qquad 3.2(12)$$

Reductive dehalogenation:

$$R—CCl_2—R \longrightarrow RCHClR + Cl^- \qquad 3.2(13)$$

Hydrolysis:

$$R—CH_2CN \longrightarrow RCHONH_3 \qquad 3.2(14)$$

An organic chemical has two levels of the biodegradation. Primary degradation refers to any biologically in-

TABLE 3.2.2   ORGANIC CHEMICALS THAT MAY RAPIDLY REACT WITH SOIL WATER AND GROUNDWATER

acetic anhydride	methacrylic acid
acetyl bromide	2-methylaziridne
acetyl chloride	methyl isocyanate
acrolein	methyl isocyanoacetate
acrylonitrile	oxopropanedinitrile
3-aminopropiononitrile	perfluorosilanes
bis(difluoroboryl)methane	peroxyacetic acid
butyldichloroborane	peroxyformic acid
calcium cyanamide	peroxyfuroic acid
2-chloroethylamine	peroxpivalic acid
chlorosulfonyl isocyanate	peroxytrifluoriacetic acid
chlorotrimethylsilane	phenylphosphonyl dichloride
cyanamide	phosphorus tricyanide
2-cyanoethanol	pivaloyloxydiethylborane
cyanoformyl chloride	potassium bis(propynyl)palladate
cyanogen chloride	potassium bis(propynyl)platinate
dichlorodimethylsilane	potassium diethynylplatinate
dichlorophenylborane	potassium hexaethynylcobaltate
dicyanoacetylene	potassium methanediazoate
diethylmagnesium	potassium tert-butoxide
diethylzinc	potassium tetracyanotitanate
diketene	potassium tetraethynylnickelate
dimethylaluminum chloride	propenoic acid
dimethylmagnesium	sulfur trixoide-dimethylformamide
dimethylzinc	sulfinylcyanamide
diphenylmagnesiuim	2,4,6-trichloro-1,3,5-triazine
2,3-epoxypropionaldehyde oxime	trichlorovinylsilane
N-ethyl-N-propylcarbamolyl chloride	triethoxydialuminum tribromide
glyoxal	vinyl acetate
isopropylisocyanide dichloride	
maleic anhydride	

Source: J. Dragun, 1988, *The soil chemistry of hazardous materials* (Silver Springs, MD: Hazardous Materials Control Research Institute).

$$k \cdot C = -\frac{dC}{dt} \qquad 3.2(15)$$

or

$$k = \frac{2.303}{t} \log\left(\frac{C_0}{C}\right) \qquad 3.2(16)$$

where:

k = rate constant, 1/time
t = time
C_0 = initial concentration, ppm
C = concentration at time t, ppm

The time needed for half of the concentration to react, half-life, can be calculated if k is known with use of the following equation:

$$t_{1/2} = \frac{0.693}{k} \qquad 3.2(17)$$

where $t_{1/2}$ is equal to half-life.

Table 3.2.3 lists the biodegradation rates for many pesticides; the biodegradation rates for other organic chemicals are in Dragun (1988b). However, note that the estimate of biodegradation rates of organic chemicals may not be accurate. Biodegradation rates can be affected by many factors such as pH, temperature, water content, carbon content, clay content, oxygen, nutrients, microbial population, acclimation, and concentration. Most of these factors are interrelated. For example, the pH can affect both the availability of a substrate as well as the composition of the microbial community.

After the degradation rate constant k is estimated, the behavior of a specific compound can be modeled with use of a form of the advection–dispersion equation, Equation 3.3(1), that includes a first-order degradation term.

duced structural alteration in the organic chemical. Ultimate biodegradation refers to the degradation of the organic chemical into carbon dioxide, oxygen, water, and other inorganic products.

Primary biodegradation of an organic chemical can generate a variety of degradation products that can contaminate groundwater. For example, the degradation of trichloroethylene (TCE) can lead to dichloroethylenes (DCEs), dichloroethanes (DCAs), vinyl chloride, and chloroethane (Dragun 1988b; Alexander 1981; Goring and Hamaker 1972). The degradation of cyclic hydrocarbons can lead to aliphatic hydrocarbons, and aliphatic hydrocarbons can be converted in successive reactions into alcohols, aldehydes, and then aliphatic acids (Tabak et al. 1981).

The biodegradation of many organic chemicals is generally first-order with respect to the organic chemical's concentration (Scow 1982). As a result, the biodegradation rate constant can be calculated with use of the following first-order equation as in hydrolysis:

ADSORPTION

Adsorption is the bonding of an organic chemical to the soil mineral surfaces (clay) or to the organic matter surfaces. The bonding is usually temporary and is accomplished by ionic, ligand, dipole, hydrogen, or Van der Waal's bonds. Adsorption is important in the movement of organic chemicals in groundwater because it decreases the mobility and retards the migration of an organic chemical in groundwater. Furthermore, the adsorbed portion of an organic chemical may not be available in solution for other chemical reactions such as hydrolysis and biodegradation.

The degree and extent of adsorption of an organic chemical to soil is determined by the chemical's structure and the soil's physical and chemical characteristics. Organic chemicals with large molecular structures, such as PCBs, PAHs, toluene, and dichlorodiphenyl trichloroethane (DDT), tend to be extensively adsorbed onto soil (Landrum et al. 1984). Organic chemicals with positive charges, such as the herbicides paraquat and diquat, are readily adsorbed onto the cation exchange sites (clay min-

TABLE 3.2.3 BIODEGRADATION RATE CONSTANTS FOR ORGANIC COMPOUNDS IN SOIL

Compound	$k \, (Day^{-1})$
Aldrin, Dieldrin	0.013
Atrazine	0.019
Bromacil	0.0077
Carbaryl	0.037
Carbofuran	0.047
DDT	0.00013
Diazinon	0.023
Dicamba	0.022
Diphenamid	0.123[b]
Fonofos	0.012
Glyphosate	0.10
Heptachlor	0.011
Lindane	0.0026
Linuron	0.0096
Malathion	1.4
Methyl parathion	0.16
Paraquat	0.0016
Phorate	0.0084
Picloram	0.0073
Simazine	0.014
TCA	0.059
Terbacil	0.015
Trifluralin	0.008
2,4-D	0.066
2,4,5-T	0.035

Source: W. Mabey and T. Mill, 1978, Critical review of hydrolysis of organic compounds in water under environmental conditions, *Jour. Phys. Chem. Ref. Data* 17, no. 2:383–415.

eral surfaces). In addition, the adsorption of organic chemicals depends on the organic matter content of the soil. The relationship between the organic content of soil and the adsorption coefficient of organic chemicals is generally linear for soils with an organic carbon content greater than 0.1 (Hamaker and Thompson 1972).

The adsorption process is usually reversible. At equilibrium, the adsorption coefficient, which is the rate at which the dissolved organic chemical in water transfers into the soil, can be described with the linear Freundlich isotherm equation as

$$K_d = \frac{C_s}{C_w} \qquad 3.2(18)$$

where:

K_d = distribution coefficient
C_s = concentration adsorbed on soil surfaces, ug/g
C_w = concentration in water, ug/ml

Other nonlinear isotherm equations are also used (Lyman 1982), such as:

$$K_d = \frac{C_s}{C_w^{1/n}} \qquad 3.2(19)$$

where n is a constant usually between 0.7 and 1.1.

As in Equations 3.2(18) and 3.2(19), the distribution or adsorption coefficient K_d is directly proportional to the organic carbon content of the soil; thus, K_d can be written as

$$K_d = \frac{K_{oc}}{f_{oc}} \qquad 3.2(20)$$

where:

K_{oc} = normalized adsorption coefficient
f_{oc} = soil organic carbon content

The normalized adsorption coefficient can be estimated from the organic chemical's water solubility or octanol water partition coefficient with use of regression equations (Dragun 1988b), such as

$$\log(K_{oc}) = a \cdot \log(S) + b \qquad 3.2(21)$$

$$\log(K_{oc}) = c \cdot \log(K_{ow}) + d \qquad 3.2(22)$$

where:

S = water solubility
K_{ow} = octanol water partition coefficient
a,b,c,d = coefficients that depend on the organic chemical

Table 3.2.4 lists the adsorption coefficient K_{oc} for several organic chemicals. The regression coefficients a, b, c, and d for several chemicals are in Brown and Flagg (1981), Briggs (1973), and Keneya and Goring (1980). Therefore, after the distribution coefficient K_d is estimated, the effect of adsorption on the mobility of a specific compound can be calculated with use of a form of the advection–dispersion equation, Equations 3.3(1) and 3.3(3), that includes the retardation factor R.

VOLATILIZATION

Volatilization is the loss of chemicals in vapor form from the soil water (liquid phase) or the soil surfaces (solid phase) to the soil air (gas phase) of the unsaturated zone. Only the first type of volatilization, from the liquid phase to the gas phase, is discussed in this section. The volatilization from the solid phase to the gas phase is relatively small and usually neglected. However, information on this type of volatilization is presented by Mayer, Letey, and Farmer (1974); Baker and Mackay (1985); and Jury, Farmer, and Spencer (1984).

The extent of volatilization of an organic chemical from water to the soil air can be determined by Henry's law which states that when a solution becomes dilute, the vapor pressure of a chemical is proportional to its concentration (Thomas 1982) as

$$C_a = H \cdot C_w \qquad 3.2(23)$$

TABLE 3.2.4 MEASURED K_{OC} VALUES FOR VARIOUS ORGANIC CHEMICALS

Chemical	K_{oc}	Chemical	K_{oc}
acetophenone	35	ipazine	1,660
alachlor	190	isocil	130
aldrin	410	isopropalin	75,250
ametryn	392	leptophos	9,300
6-aminochrysene	162,900	linuron	813
anthracene	26,000	malathion	1,778
asulam	300	methazole	2,620
atrazine	148	methomyl	160
benefin	10,700	methoxychlor	80,000
alpha-BHC	1995	methylparathion	5,129
beta-BHC	1995	metobromuron	60
2,2′-biquinoline	10,471	metribuzin	95
bromacil	72	monolinuron	200
butraline	8,200	monuron	100
carbaryl	229	napthalene	1,300
carbofuran	105	napropamide	680
carbophenothion	45,400	neburon	2,300
chloramben	21	nitralin	960
chlorobromuron	460	nitrapyrin	458
chloroneb	1159	norflurazon	1,914
chloroxuron	3200	oxadiazon	3,241
chlorpropham	589	parathion	4,786
chlopyrifos	13,490	pebulate	630
crotoxyphos	170	phenathrene	23,000
cyanazine	200	phenol	27
cycloate	345	phorate	3,200
2,4-D	57	picloram	17
DBCP	129	profluralin	8,600
p,p′-DDT	129	prometon	350
diallate	1,900	prometryn	48
diamidaphos	32	pronamide	200
dicamba	0.4	propachlor	265
dichlobenil	235	propazine	158
dinitramine	4,000	propham	51
dinoseb	124	pyrazon	120
dipropetryn	1,170	pyrene	62,700
disulfoton	1,780	pyroxychlor	3,000
diruon	398	silvex	2,600
DMSA	770	simazine	135
EPTC	240	2,4,5-T	53
ethion	15,400	tebuthiuron	620
fenuron	27	terbacil	51

Source: J. Dragun, 1988, *The soil chemistry of hazardous materials* (Silver Springs, MD: Hazardous Materials Control Research Institute).

where:

C_a = concentration of the chemical in air
H = Henry's law constant
C_w = concentration of the chemical in water

Henry's law constant for a chemical can be calculated with the following equation:

$$H = \frac{P_v \cdot M_w}{760 \cdot S} \qquad 3.2(24)$$

where:

P_v = vapor pressure of the chemical in mmH$_g$
M_w = molecular weight of the chemical
S = solubility in mg/l

Published texts report Henry's law constant in various units such as atm-m^3/mole, atm-cm^3/g, or dimensionless depending on the units used for C_a and C_w. Table 3.2.5 lists Henry's constants for several organic chemicals. According to Lyman and others (1982), if H is less than 10^{-7} atm-m^3/mole, the substance has a low volatility. If H is less than 10^{-5} but greater than 10^{-7} atm-m^3/mole, the substance volatilizes slowly. However, the volatilization becomes an important transfer mechanism when H is greater than 10^{-5} atm-m^3/mole.

Several soil characteristics affect the volatilization of organic chemicals in groundwater. Volatilization decreases as the soil porosity decreases or as the soil water content increases. Soils with high clay content tend to have a high water content and hence low volatilization (Jury 1986).

Inorganic Contaminants

Comprehensive information on the behavior of most inorganic chemicals in groundwater is limited. Agriculturally important compounds have been studied for many years; however, inorganic compounds such as metals have only recently begun to attract widespread interest as groundwater and soil contamination become a concern. This section illustrates some of the most important processes for several groups of inorganic contaminants and the impact of these processes on the concentration and mobility of contaminants.

Inorganic constituents in the subsurface environment can be classified into the following four categories: nutrients, acids and bases, halides, and metals. The origin and sources of these inorganics are discussed in Section 3.1.

NUTRIENTS

Nutrients such as nitrogen, phosphorous, and sulfur are essential for plant and microorganism growth. They are either applied to the land surface to increase its fertility or discarded with waste streams that contain appreciable amounts of these nutrients. These nutrients, however, can have appreciable concentrations that can leach into the ground and adversely affect the quality of groundwater.

Nitrogen (N) is found in waste, soil, and the atmosphere in various forms such as ammonia, ammonium, nitrite, nitrate, and molecular nitrogen. Nitrogen is converted to ammonium (NH_4^+) by a process called *ammonification*. Because of its positive charge, ammonium can be held in the soil on cation exchange sites. Ammonium can also be converted temporarily to nitrite (NO_2^-) and then to nitrate (NO_3^-) by aerobic nitrifying organisms through a process called *nitrification*. Ammonification and nitrification normally occur in the unsaturated zone where microorgan-

TABLE 3.2.5 VALUES OF HENRY'S LAW CONSTANT FOR SELECTED CHEMICALS

Low Volatility ($H < 3 \times 10^{-7}$)	H attm-m³	H' (non-dim.)	High Volatility ($H < 10^{-3}$)	H attm-m³	H' (non-dim.)
3-Bromo-1-propanol	1.1×10^{-7}	4.6×10^{-6}	Ethylene dichloride	1.1×10^{-3}	4×10^{-2}
Diedrin	2×10^{-7}	8.9×10^{-6}	Naphthalene	1.15×10^{-3}	4.9×10^{-2}
			Biphenyl	1.5×10^{-3}	6.8×10^{-2}
Middle Range ($3 \times 10^{-7} < H < 10^{-3}$)			Aroctor 1254	2.7×10^{-3}	1.6×10^{-1}
			Methylene chloride	3×10^{-3}	1.3×10^{-1}
			Aroctor 1248	3.5×10^{-3}	1.6×10^{-1}
Lindane	4.8×10^{-7}	2.2×10^{-5}	Chlorobenzene	3.7×10^{-3}	1.65×10^{1}
m-Bromonitrobenzene	1.6×10^{-6}	7.4×10^{-5}	Chloroform	4.7×10^{-3}	2.0×10^{-1}
Pentachlorophenol	3.4×10^{-6}	1.5×10^{-4}	o-Xylene	5.1×10^{-3}	2.2×10^{-1}
4-tert-Butylphenol	9.1×10^{-6}	3.8×10^{-4}	Benzene	5.5×10^{-3}	2.4×10^{-1}
Triethylamine	1.3×10^{-5}	5.4×10^{-4}	Toluene	6.6×10^{-3}	2.8×10^{-1}
Aldrin	1.4×10^{-5}	6.1×10^{-4}	Aroclor 1260	7.1×10^{-3}	3.0×10^{-1}
Nitrobenzene	2.2×10^{-5}	9.3×10^{-4}	Perchloroethylene	8.3×10^{-3}	3.4×10^{-1}
Epichlorohydrin	3.2×10^{-5}	1.3×10^{-3}	Ethyl benzene	8.7×10^{-3}	3.7×10^{-1}
DDT	3.8×10^{-5}	1.7×10^{-3}	Trichloroethylene	1×10^{-2}	4.2×10^{-1}
Phenanthrene	3.9×10^{-5}	1.7×10^{-3}	Mercury	1.1×10^{-2}	4.8×10^{-1}
Acenaphthene	1.5×10^{-4}	6.2×10^{-3}	Methyl bromide	1.3×10^{-2}	5.6×10^{-1}
Acetylene tetrabromide	2.1×10^{-4}	8.9×10^{-3}	Cumene (isopropyl)	1.5×10^{-2}	6.2×10^{-1}
Aroclor 1242	5.6×10^{-4}	2.4×10^{-2}	1,1,1-Trichloroethane	1.8×10^{-2}	7.7×10^{-1}
Ethylene dibromide	6.6×10^{-4}	2.8×10^{-2}	Carbon tetrachloride	2.3×10^{-2}	9.7×10^{-1}
			Methyl chloride	2.4×10^{-2}	9.7×10^{-1}
			Ethyl bromide	7.3×10^{-2}	3.1
			Vinyl chloride	2.4	99
			2,2,4-Trimethyl pentane	3.1	129
			n-Octane	3.2	136
			Fluorotrichloromethane	5.0	—
			Ethylene	>8.6	~360

Source: R.G. Thomas, 1982, Volatilization from water, In *Handbook of chemical property estimation methods* (New York: McGraw-Hill, Inc.).

isms and oxygen are abundant, but nitrate can be readily leached from the soil into groundwater where it may present a health hazard; nitrate is highly mobile in groundwater because of its negative charge. *Denitrification* is a process whereby NO_3^- is reduced to nitrous oxide (N_2O) and elemental nitrogen (N_2) by facultative anaerobic bacteria (Downing, Painter, and Knowles 1964; Freeze and Cherry 1979; Bemner and Shaw 1958).

Phosphorous (P) is found in organic waste, rock phosphate quarries, fertilizers, and pesticides in concentrations high enough to potentially leach into groundwater. The decomposition of organic waste and dissolution of inorganic fertilizers provide soluble phosphorous, soluble orthophosphate, and a variety of condensed phosphates, tripolyphosphates, adsorbed phosphates, and crystallized phosphates (U.S. EPA 1983). The hydrolysis and mineralization of these products provide soluble phosphate which can be used by plants and microorganisms, adsorbed to soil particles, or leached to groundwater. Although phosphorous is not a harmful constituent in drinking water, its presence in groundwater is environmentally significant if the groundwater discharges to a surface water body where phosphorous can produce algae growth and cause eu-

trophication of the aquatic system (Freeze and Cherry 1979).

Sulfur (S) is found in appreciable amounts in waste streams from kraft mills, sugar refining, petroleum refining, and copper and iron extraction facilities (Overcash and Pal 1979). Aerobic bacteria can oxidize the reduced forms of sulfur to form sulfate which can be highly adsorbed to soil when the cation adsorbed on the clay is aluminum; moderately adsorbed when the cation is calcium; and weakly adsorbed when the cation is potassium (Tisdale and Nelson 1975). Leaching losses of sulfur to groundwater can be large because of the anionic structure of sulfur and the solubility of most of its salt. Leaching is greatest when monovalent cations such as potassium and sodium predominate; moderate when calcium and manganese predominate; and minimal when the soil is acidic and appreciable levels of exchangeable iron and aluminum are present (Tisdale and Nelson 1975).

ACIDS AND BASES

Industrial liquid wastes are comprised of large volumes of inorganic acids and bases that can alter the soil's properties. Acids can increase the amount of aluminum (Al), iron (Fe), and other cations in the water phase of the soil system as the hydrogen ion (H^+) cation competes for cation exchange sites. If significant amounts of H^+ are present, they can dissolve the more acid-solid minerals, releasing cations which are previously fixed to the mineral structure into the water phase (Dragun 1988b). In addition, acids can cause the dissolution of some of the clay minerals and generally increase soil permeability. Bases can increase the amount of cations in the water phase by dissolving the more base-soluble soil minerals. Bases can also cause the dissolution of some of the soil's predominant clay minerals and generally decrease soil permeability.

HALIDES

Halides are the stable anions of the highly reactive halogens: fluoride (F), chloride (Cl), bromine (Br), and iodine (I). Halides occur naturally in soils and are also present in many industrial waste streams.

Fluoride is present in phosphatic fertilizers, hydrogen fluoride, fluorinated hydrocarbons, and certain petroleum refinery waste. The leaching losses and mobility of fluoride can be large because of the anionic structure of fluoride and the solubility of some of its salt (Bemner and Shaw 1958). Sodium salts of fluoride (NaF) are soluble and result in high soluble fluoride levels in soils low in calcium. Calcium salts of fluoride (CaF_2), however, are relatively insoluble and limit the amount of fluoride leached to groundwater. Fluoride solubility depends on the kind and relative quantity of cations present in soil that has formed salts with the fluoride ion (F^-). Fluorosis disease can occur in animals who consume water containing 15 ppm of fluoride (Lee 1975).

Chloride (Cl) is present in chlorinated hydrocarbon production and chlorine gas production wastes as well as other wastes. Chloride is soluble and mobile in groundwater because of its anionic structure.

Bromide (Br) is present in synthetic organic dyes, mixed petrochemical wastes, photographic supplies, and pharmaceutical and inorganic wastes. Other forms of bromide such as bromate and bromic acid occur naturally in soils at smaller concentrations. Most bromide salts (CaBr, MgBr, NaBr, and Kbr) are soluble and readily leachable into water percolating through the soil and down to groundwater (U.S. EPA 1983).

Iodine (I) is present in pharmaceutical and chemical industrial wastes. Iodine is only slightly water soluble and tends to be retained in soils by forming complexes with organic matter and being fixed to phosphates and sulfates.

METALS

Metals are found in industrial wastes in a variety of forms. When these metals are introduced into the subsurface environment, they can react with water and soil in several physicochemical processes to produce appreciable concentrations that affect the quality of groundwater. The most important processes that affect the concentration and mobility of metals in groundwater include filtration, precipitation, complexation, and ion exchange.

Filtration occurs when dissolved and solid matter are trapped in the pore spaces clogging the pore spaces and decreasing the permeability of the soil system (Dragun 1988b).

Precipitation occurs when metal ions react with water to form reaction products which precipitate in soil as oxide and oxyhydroxide minerals or form oxyde and oxyhydroxide coatings on soil minerals. Precipitation of metals as hydroxides, sulfides, and carbonates is common (Dragun 1988b).

Complexation involves the formation of soluble, charged or neutral complexes between metal ions and inorganic or organic anions called *ligands*. The complexes formed influence the mobility and concentration of the metal in groundwater. For example, the mobility of zinc in groundwater is affected by the formation of complex species between the zinc ion and inorganic anions present in the water, such as HCO_3^-, CO_3^{2-}, SO_4^{2-}, Cl^-, F^-, and NO_3^- (Freeze and Cherry 1979). The complexation of cobalt-60 ions by synthetic organic compounds enhances its mobility in groundwater (Killey et al. 1984). Other metal species are reported to be highly mobile in groundwater after soluble complexes are formed with humic substances or organic solvents (Bradbent and Ott 1957; Griffin and Chou 1980).

The predominant complex species in an aqueous solution are influenced by the redox and pH of the soil. The relationship between the redox, pH, and the complex species is commonly expressed in Eh–pH diagrams for each metal; Eh is the electronic potential. Figure 3.2.1 shows an example of an Eh–pH diagram for mercury. Methods for calculating Eh–pH diagrams are discussed by several authors (Brookings 1980; Garrells and Christ 1965; Verink 1979).

Using Eh–pH diagrams, environmental engineers can qualitatively determine the most important complexes formed by the metal in water and estimate the concentration and mobility of the metal in groundwater. The concentration of cations reported in chemical analyses of groundwater normally represents the total concentration of each element in water. However, most cations exist in more than one molecular or ionic form. These forms can have different valences and, therefore, different mobilities due to different affinities to sorption and different solubility controls.

Adsorption is another process affecting the concentration and mobility of metals in groundwater. Positive adsorption involves the attraction of metal cations in water by negatively charged soil particles. Therefore, adsorption can decrease the concentration of dissolved metals in water and retard their movement. The cation exchange ca-

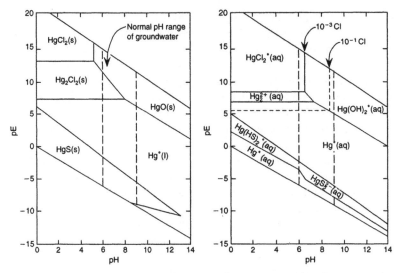

FIG. 3.2.1 Stability fields of solid phases and aqueous species of mercury as a function of pH and Eh at 1 bar total pressure. (Reprinted from J.D. Hem, 1967, Equilibrium chemistry of iron in groundwater, In *Principles and applications of water chemistry*, edited by S.D. Faust and J.V. Hunter, New York: John Wiley and Sons.)

pacity (CEC) of a soil, defined as the amount of cations adsorbed by the soil's negative charges, is usually expressed as milliequivalents (meq) per 100 grams of soil. In general, clay soils and humus have a higher CEC than other soils.

Some cations are more attracted to a soil surface than others based on the size and charge of their molecule. For example the Cu^{2+} cation in water can displace and replace a Ca^{2+} cation present at the soil surface through a process known as ion exchange. Also, trivalent cations are preferentially adsorbed over divalent cations which are preferentially adsorbed over monovalent cations. The release of

ions by exchange processes can aggravate a contamination problem. For example, increases in water hardness resulting from the displacement of calcium and magnesium ions from geological materials by sodium or potassium in landfill leachate have been documented (Hughes, Candon, and Farvolden 1971).

The cation exchange is reversible, and its extent can be described by the adsorption or distribution coefficient (Dragun 1988b) as

$$K_d = \frac{C_s}{C_w} \qquad 3.2(25)$$

where:

K_d = adsorption or distribution coefficient
C_s = concentration adsorbed on soil surfaces (ug/g of soil)
C_w = concentration in water (ug/ml)

Table 3.2.6 lists the adsorption coefficients for several metals. The greater the coefficient K_d, the greater the extent of adsorption. Furthermore, changes in metal concentration, as well as pH, can have a significant effect on the extent of adsorption as shown in Figure 3.2.2.

A negative adsorption occurs when anions (negatively charged metal ions) are repulsed by negative soil particle charges. This repulsion causes high mobility and migration of anions in water. This process is also known as *anion exclusion*.

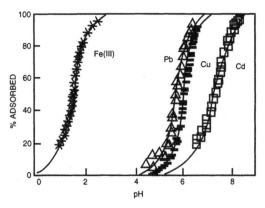

FIG. 3.2.2 Adsorption of metal ions on amorphous silica as a function of pH. (Reprinted from U.S. Environmental Protection Agency, 1989, *Transport and fate of contaminants in the subsurface*, Seminar Publication EPA/625/4-89/019, Cincinnati: U.S. EPA.)

—Ahmed Hamidi

TABLE 3.2.6 RANGES FOR K_D FOR VARIOUS ELEMENTS IN SOILS AND CLAYS

Element	Observed Range (ml/g)	Mean[a]	Standard Deviation[b]
Ag	10–1,000	4.7	1.3
Am	1.0–47,230	6.7	3.0
As(III)	1.0–8.3	1.2	0.6
As(V)	1.9–18	1.9	0.5
Ca	1.2–9.8	1.4	0.8
Cd	1.3–27	1.9	0.9
Ce	58–6,000	7.0	1.3
Cm	93–51,900	8.1	1.9
Co	0.2–3,800	4.0	2.3
Cr(III)	470–150,000	7.7	1.2
Cr(VI)	1.2–1,800	3.6	2.2
Cs	10–52,000	7.0	1.9
Cu	1.4–333	3.1	1.1
Fe	1.4–1,000	4.0	1.7
K	2.0–9.0	1.7	0.5
Mg	1.6–13.5	1.7	0.5
Mn	0.2–10,000	5.0	2.7
Mo	0.4–400	3.0	2.1
Np	0.2–929	2.4	2.3
Pb	4.5–7,640	4.6	1.7
Po	196–1,063	6.3	0.7
Pu	11–300,000	7.5	2.3
Ru	48–1,000	6.4	1.0
Se(IV)	1.2–8.6	1.0	0.7
Sr	0.2–3,300	3.3	2.0
Te	0.003–0.28	3.4	1.1
Th	2,000–510,000	11.0	1.5
U	11–4,400	3.8	1.3
Zn	0.1–8,000	2.8	1.9

Source: C.F. Baes III and R.D. Sharp, 1983, A proposal for estimation of soil leaching and leaching constants for use in assessment models, *Journal of Environmental Quality* 12, no. 1:17–28.

[a]Mean of the logarithms of the observed values.
[b]Standard deviation of the logarithms of the observed values.

References

Alexander, M. 1981. Biodegradation of chemicals of environmental concern. *Science* 211:132–138.

Arthur D. Little, Inc. 1976. *Physical, chemical, and biological treatment techniques for industrial wastes.* Report to U.S. EPA, Office of Solid Waste Management Programs, PB-275-054/56A Vol. 1 and PB-275-278/1GA Vol. 2.

Baker, L.W., and K.P. Mackay. 1985. Screening models for estimating toxic air pollution near a hazardous waste landfill. *Journal of the Air Pollution Control Association* 35, no. 11:1190–1195.

Bemner, J.M., and K. Shaw. 1958. Denitrification in soils-factors affecting denitrification. *Journal of Agricultural Science* 51, no. 1:22–52.

Bradbent, F.E., and J.B. Ott. 1957. Soil organic matter: Metal complexes—factors affecting various cations. *Soil Science* 83:419–427.

Briggs, G.G. 1973. A simple relationship between soil adsorption of organic chemicals and their octanaol/water partition coefficients. Proceedings of the 7th British Insecticide and Fungicide Conference. Vol. 1. Nottingham, Great Britain: The Boots Company, Ltd.

Brookings, D.G. 1980. *Eh–pH diagrams for elements of interest at the Oklo Natural Reactor at 25°C, 1 bar pressure and 200°C, 1 bar pressure.* Report to Los Alamos National Laboratory, CNC-11.

Brown, D.S., and E.W. Flagg. 1981. Empirical prediction of organic pollutant adsorption in natural sediments. *Journal of Environmental Quality* 10:382–386.

Burkhard, N., and J.A. Guth. 1981. Chemical hydrolysis of 2-chloro-4, 6-bis (alkylamino)-1,3,5-triazine herbicides and their breakdown in soils under the influence of adsorption. *Pestic. Sci* 12:45–52.

Downing, A.L., H.A. Painter, and C. Knowles. 1964. Nitrification in the activated sludge process. *Journal and Proceedings of the Institute of Sewage Purification,* Part 2.

Dragun, J. 1988a. Microbial degradation of petroleum products in soil. Proceedings of a Conference on Environmental and Public Health, Effects of Soils Contaminated with Petroleum Products, October, New York: John Wiley and Sons.

———. 1988b. *The soil chemistry of hazardous materials.* Silver Springs, Md.: Hazardous Materials Control Research Institute.

Dragun, J., and C.S. Helling. 1985. Physicochemical and structural relationships of organic chemicals undergoing soil and clay catalyzed free-radical oxidation. *Soil Science* 139:100–111.

Freeze, R.A., and J.A. Cherry. 1979. *Groundwater.* Englewood Cliffs, N.J.: Prentice-Hall.

Garrells, R.M., and C.L. Christ. 1965. *Minerals, solutions, and equilibria.* New York: Harper and Row.

Goring, C.A.J., and J.W. Hamaker. 1972. *Organic chemicals in the soil environment.* Vols. 1 and 2. New York: Marcel Dekker.

Griffin, R.A., and S.F.J. Chou. 1980. *Attenuation of polybrominated biphenyls and hexachlorobenzene by earth materials.* Environmental Geology Notes 87, Illinois State Geological Survey. Urbana, Ill.

Hamaker, J.W., and J.M. Thompson. 1972. Adsorption. In *Organic chemicals in the soil environment,* edited by C.A.I. Goring and J.W. Hamaker. New York: Marcel Dekker.

Harris, J.C. 1982. Rate of hydrolysis. In *Handbook of chemical property estimation methods.* New York: McGraw-Hill.

Hughes, G.M., R.A. Candon, and R.N. Farvolden. 1971. *Hydrogeology of solid waste disposal sites in northern Illinois.* Solid Waste Management Series, report SW-124. U.S. EPA.

Jury, W.A. 1986. Volatilization from soil. In *Vadoze modeling of organic contaminants,* edited by Stephen Hern and Susan Melancon, 159–176. Chelsea, Mich.: Lewis Publishers.

Jury, W.A., W.J. Farmer, and W.F. Spencer. 1984. Behavior assessment model for trace organics in soils. *Journal of Environmental Quality* 13, no. 4.

Keneya, E.E., and C.A.I. Goring. 1980. Relationship between water solubility, soil-sorption, octanol-water partitioning, and bioconcentration of chemicals in biota. In *Aquatic toxicology,* ASTM STP 707. Philadelphia, Pa.: ASTM.

Killey, R.W. et al. 1984. Subsurface cobalt-60 migration from a low-level waste disposal site. *Environmental Science Technology* 18, no. 3:148–156.

Konrad, J.G., and G. Chesters. 1969. Degradation in soils of ciodrin and organophosphate insecticide. *J. Agr. Food Chem.* 17:226–230.

Landrum, P.F. et al. 1984. Reverse-phase separation method for determining pollutant binding to aldrich humic acid and dissolved organic carbon of natural waters. *Environmental Science and Technology* 18:187–192.

Lee, H.L. 1975. Trace elements in animal production. In *Trace elements in soil-plant-animal systems,* edited by D. Nicholas and R. Egan. New York: Academic Press.

Lyman, W. 1982. Adsorption coefficient for soils and sediments. In *Handbook of chemical property estimation.* New York: McGraw-Hill.

Lyman, W.J., W.F. Reehl, and D.H. Rosenblatt. 1982. *Handbook of chemical property estimation methods: Environmental behavior of organic compounds.* New York: McGraw-Hill.

Mayer, R., J. Letey, and W.J. Farmer. 1974. Models for predicting volatilization of soil-incorporated pesticides. Soil Science Society of America Proceedings. Vol. 38:563–568.

Mabey, W., and T. Mill. 1978. Critical review of hydrolysis of organic compounds in water under environmental conditions. *Jour. Phys. Chem. Ref. Data* 17, no. 2:383–415.

Overcash, M.R., and D. Pal. 1979. *Design of land treatment systems for industrial wastes; Theory and practice.* Ann Arbor, Mich.: Ann Arbor Science.

Rao, P.S.C., and R.E. Jessup. 1982. Development and verification of simulation models for describing pesticide dynamics in soils. *Ecol. Modeling* 16:67–75.

Sax, N.I. 1984. *Dangerous properties of industrial materials.* 6th ed. New York: Van Nostrand Reinhold.

Scow, K.M. 1982. Rate of biodegradation. In *Handbook of chemical property estimation methods.* New York: McGraw-Hill.

Tabak, H.N. et al. 1981. Biodegradability studies with organic priority pollutants compounds. *Journal Water Pollution Control Federation* 53:1503–1518.

Thomas, R.G. 1982. Volatilization from water. In *Handbook of chemical property estimation methods.* New York: McGraw-Hill.

Tisdale, S.L., and W.L. Nelson. 1975. *Soil fertility and fertilizers.* 3d ed. New York: Macmillan.

U.S. Environmental Protection Agency 1983. *Hazardous waste land treatment.* SW-874. Washington, D.C.: U.S. EPA, Office of Solid Waste and Emergency Response.

Valentine, R.L. 1986. *Vadoze zone modeling of organic pollutants,* edited by Stephen Hern and Susan Melancon, 233–243. Chelsea, Mich.: Lewis Publishers.

Valentine, R.L., and J.L. Schnoor. 1986. Biotransformation. In *Vadoze zone modeling of organic pollutants,* edited by Stephen Hern and Susan Melancon. Chelsea, Mich.: Lewis Publishers.

Verink, E.D. 1979. Simplified procedure for constructing pourbaix diagrams. *Journal of Education Modules Math. Sci. Eng.* 1:535–560.

3.3
TRANSPORT OF CONTAMINANTS IN GROUNDWATER

This section discusses the transport of contaminants in groundwater and describes the transport process and the behavior of the contaminant plume.

Transport Process

When a contaminant is introduced in groundwater, it spreads and moves with the groundwater as a result of (1) advection which is caused by the flow of groundwater, (2) dispersion which is caused by mechanical mixing and molecular diffusion, and (3) retardation which is caused by adsorption. The mathematical relationship between these processes can be written as follows (Javandel, Doughtly, and Tsang 1984):

$$\frac{\partial}{\partial x_i}\left[D_{ij}\frac{\partial C}{\partial x_j}\right] - \frac{\partial}{\partial x_i}(Cv_i) - \frac{C'W'}{n} = R\frac{\partial C}{\partial t} \qquad 3.3(1)$$

$$v_i = \frac{-K_{ij}}{n}\frac{\partial h}{\partial x_j} \qquad 3.3(2)$$

$$R = \left[1 + \frac{\rho_b K_d}{n}\right] \qquad 3.3(3)$$

where:

- C = contaminant concentration
- v_i = seepage or average pore water velocity in the direction x_i
- D_{ij} = dispersion coefficient
- K_{ij} = hydraulic conductivity
- C' = solute concentration in the source or sink fluid
- W' = volume flow rate per unit volume of the source or sink

- n = effective porosity
- h = hydraulic head
- R = retardation factor
- x_i = cartesian coordinate

The following discussion uses a simplified two-dimensional representation of Equation 3.3(1) to describe the transport of contaminants in groundwater. In a homogeneous, isotropic medium having a unidirectional steady-state flow with seepage velocity V, Equation 3.3(1) can be rewritten as

$$D_L\frac{\partial^2 C}{\partial x^2} + D_T\frac{\partial^2 C}{\partial y^2} - V\frac{\partial C}{\partial x} = R\frac{\partial C}{\partial t} \qquad 3.3(4)$$

where:

- C = contaminant concentration
- V = seepage or average pore water velocity
- D_L = longitudinal dispersion coefficient
- D_T = transversal dispersion coefficient
- R = retardation factor

ADVECTION

A contaminant moves with the flow of groundwater according to Darcy's law. Darcy's law states that the flow rate of water through soil from point 1 to point 2 is proportional to the head loss and inversely proportional to the length of the flow path as

$$Q = -K \cdot A\frac{h_2 - h_1}{L} \qquad 3.3(5)$$

where:

Q = groundwater flow rate
A = cross-sectional area of flow
$h_2 - h_1$ = head loss between point 1 and point 2
L = distance between point 1 and point 2
K = hydraulic conductivity

The actual seepage or average pore water velocity can be calculated as

$$V = \frac{Q}{n \cdot A} = -\frac{K}{n} \frac{h_2 - h_1}{L} \qquad 3.3(6)$$

where n is the effective porosity or percent of interconnected pore spaces that actually contributes to the flow.

The average pore water velocity calculated in Equation 3.3(6) is a conservative estimate of the migration velocity of the contaminant in groundwater. Therefore, when only advection is considered, a contaminant moves with the groundwater flow at the same rate as water, and no diminution of concentration is observed. In reality, however, the movement of the contaminant is also influenced by dispersion and retardation.

DISPERSION

Dispersion is the result of two processes, molecular diffusion and mechanical mixing.

Molecular diffusion is the process whereby ionic or molecular constituents move under the influence of their kinetic activity in the direction of their concentration gradients. Under this process, constituents move from regions of higher concentration to regions of lower concentration; the greater the difference, the greater the diffusion rate. Molecular diffusion can be expressed by Fick's law as

$$F = -D_f \frac{dC}{dx} \qquad 3.3(7)$$

where:

F = mass flux per unit area per unit time
D_f = diffusion coefficient
C = contaminant concentration
dC/dx = concentration gradient

Fick's law was derived for chemicals in unobstructed water solutions. When this law is applied to porous media, the diffusion coefficient should be smaller because the ions follow longer paths between solid particles and because of adsorption. This application yields an apparent diffusion coefficient D* represented by

$$D^* = w \cdot D_f \qquad 3.3(8)$$

where w is an empirical coefficient less than 1. Perkins and Johnston (1963) suggest an approximate value of 0.707 for w. Bear (1979) suggests that w is equivalent to the tortuosity of the porous medium with a value close to 0.67. Values of D* for major ions can be obtained from Robinson and Stokes (1965).

Mechanical mixing is the result of velocity variations within the porous medium. The velocity is greater in the center of the pore space between particles than at the edges. As a result, the contaminant spreads gradually to occupy an ever-increasing portion of the flow field. Mechanical mixing dispersion can occur both in the longitudinal direction of the flow as well as in the transverse direction. According to Bachmat and Bear (1964), the mechanical mixing component of dispersion can be assumed proportional to the seepage velocity as

$$D_{11} = a_L \cdot V \qquad 3.3(9)$$

$$D_{22} = a_T \cdot V \qquad 3.3(10)$$

where:

D_{11} = longitudinal mechanical mixing component of dispersion
D_{22} = transversal mechanical mixing component of dispersion
a_L = longitudinal dispersivity
a_T = transversal dispersivity
V = average linear pore water velocity

Finally the hydrodynamic dispersion coefficients can be written as

$$D_L = D_{11} + D_f = a_L \cdot V + D^* \qquad 3.3(11)$$

$$D_T = D_{22} + D_f = a_T \cdot V + D^* \qquad 3.3(12)$$

The dispersivity coefficients a_L and a_T are characteristic of the porous medium. Representative values of dispersivity coefficients can be determined from breakthrough column tests in the laboratory or tracer tests in the field (Anderson 1979).

Figure 3.3.1 shows how dispersion can cause some of the contaminant to move faster than the average groundwater velocity and some of the contaminant to move slower than the average groundwater velocity. The front of the contaminant plume is no longer sharp but rather smeared. Therefore, when dispersion is also considered,

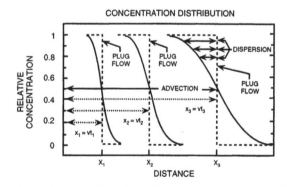

FIG. 3.3.1 Effect of dispersion–advection on concentration distribution. (Reprinted from U.S. Environmental Protection Agency, 1989, *Transport and fate of contaminants in the subsurface*, Seminar Publication EPA/625/4-89/019 (Cincinnati: U.S. EPA.)

the contaminant actually moves ahead of what would have been predicted by advection only.

RETARDATION

Retardation in the migration of contaminants in groundwater is due to the adsorption mechanism, which was described in Section 3.2 for both organic and inorganic constituents. The retardation coefficient can be calculated based on the distribution or adsorption coefficients of the contaminant and the characteristics of the porous medium as

$$R = \left[1 + K_d \frac{\rho_d}{n}\right] \qquad 3.3(13)$$

where K_d is the distribution or adsorption coefficient described previously. The values ρ_d and n are the bulk density and porosity of the soil. The velocity of the contaminant in groundwater can be calculated as follows:

$$V_c = \frac{V}{R} \qquad 3.3(14)$$

where V_c is the velocity of the contaminant movement in groundwater, V is the groundwater velocity, and R is the retardation factor. A high retardation factor, i.e., high adsorption coefficient, significantly retards the movement of the contaminant in groundwater. Figure 3.3.2 illustrates the effect of advection, dispersion, and retardation on the mobility of a contaminant in groundwater.

Contaminant Plume Behavior

The behavior and movement of contaminants in groundwater depend on the solubility and density of the contaminant, groundwater flow regime, and the local geology. This section qualitatively discusses the effect of each of these factors on the contaminant plume.

CONTAMINANT DENSITY

Immiscible fluids such as oils do not readily mix with water; therefore, they either float on top of the water table or sink into the groundwater depending on their density. Immiscible fluids with densities less than water, also called light nonaqueous phase liquids (LNAPLs) or *floaters,* form a separate phase that can float on the groundwater table. For example, if a light-bulk hydrocarbon is released from a surface spill as shown in Figure 3.3.3, it migrates downward in the unsaturated zone due to gravity and capillary forces. If the volume of the released hydrocarbon is large, the hydrocarbon reaches the groundwater and forms a pancake on top of the water table. The pancake tends to spread laterally and in the downgradient direction until it reaches residual saturation. A portion of the pancake dis-

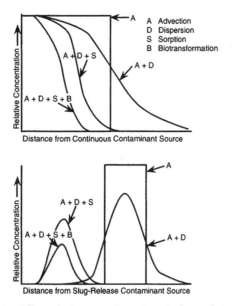

FIG. 3.3.2 Effect of advection, dispersion, and retardation on the mobility of a contaminant in groundwater. (Reprinted from M. Barcelona, 1990, *Contamination of groundwater: Prevention, assessment, restoration,* Pollution Technology Review No. 184, Park Ridge, N.J.: Nayes Data Corporation.)

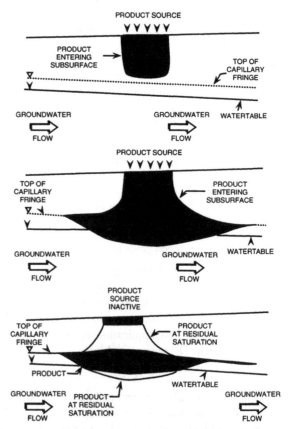

FIG. 3.3.3 Movement of LNAPLs into the subsurface. (Reprinted from U.S. Environmental Protection Agency, 1989, *Transport and fate of contaminants in the subsurface,* Seminar Publication EPA/625/4-89/019 (Cincinnati: U.S. EPA.)

solves in groundwater and eventually migrates with the water. The maximum spread of the pancake over the groundwater table can be estimated (CONCAWE Secretariat 1974) by

$$S = \frac{1000}{F}\left[V - \frac{A \cdot D}{K}\right]$$ 3.3(15)

where:

S = maximum spread of the pancake, m^2
F = thickness of the pancake, mm
V = volume of infiltrating bulk hydrocarbon, m^3
A = area of infiltration, m^2
d = depth to groundwater, m
K = constant dependent on the soil's retention capacity for oil

Table 3.3.1 lists K values for different types of hydrocarbons and soil textures.

Immiscible fluids with densities greater than water, also called dense nonaqueous phase liquids (DNAPL) or *sinkers,* sink through the saturated zone and show a concentration gradient through the aquifer, becoming more concentrated near the aquifer base as shown in Figure 3.3.4. Fingering of the dense fluid into the water can also occur depending on the characteristics of the aquifer and the viscosity of the fluid (Dragun 1988). The downward migration of the sinker can continue until a zone of lower permeability, such as a clay confining layer or a bedrock surface, is encountered. Halogenated hydrocarbons and coal tars are the principal solvents possessing densities greater than that of water. Examples of DNAPLs include methylene chloride, chloroform, trichloroethylene (TCE), tetrachloroethylene or perchloroethylene (PCE), and various Freons.

Another important factor of both LNAPL and DNAPL plume behavior is residual contamination. As the plume migrates downward through the unsaturated or saturated zone, a small amount of fluid remains attached to soil particles and within the soil pore spaces via capillarity forces. This residual contamination can reside in the soil for many

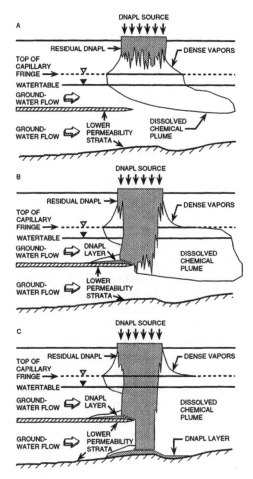

FIG. 3.3.4 Movement of DNAPLs into the subsurface. (Reprinted from S. Fenestra and J.A. Cherry, Dense organic solvents in groundwater: An introduction, In *Dense chlorinated solvents in groundwater*, Progress Report 0863985 (Ontario, Canada: Institute of Groundwater Research, University of Waterloo.)

years and serve as a continuous source of contamination. For more information on density flow, see Schwille (1988) and Fenestra and Cherry (1988).

CONTAMINANT SOLUBILITY

The solubility of a substance in water is defined as the saturated concentration of the substance in water at a given temperature and pressure. This parameter is important in the prediction of a contaminant plume in groundwater and in planning for its recovery. Substances with high water solubility have a tendency to remain dissolved in the water column, not adsorbed onto soil particles, and are more susceptible to biodegradation. Conversely, substances with low water solubility tend to adsorb onto soil particles and volatilize more readily from water. The water solubility of several substances is listed in Montgomery (1989). Several compounds, such as bulk hydrocarbons, are comprised of numerous individual chemicals and substances with different solubilities in water and different adsorption coeffi-

TABLE 3.3.1 TYPICAL VALUES FOR K FOR VARIOUS SOIL TEXTURES

	K		
Soil Texture	*Gasoline*	*Kerosene*	*Light Fuel Oil*
Stone & Coarse Gravel	400	200	100
Gravel & Coarse Sand	250	125	62
Coarse & Medium Sand	130	66	33
Medium & Fine Sand	80	40	20
Fine Sand & Silt	50	25	12

Sources: CONCAWE Secretariat, 1974, *Inland oil spill clean-up manual,* Report no. 4/74 (The Hague, The Netherlands: CONCAWE). D.N. Dietz, 1970, Pollution of permeable strata by oil components, In *Water pollution by oil,* edited by P. Hepple (New York: Elsevier Publishing and Institute of Petroleum).

cients in soil. When these compounds are introduced in groundwater, they generate contaminant plumes with different shapes and rates of migration.

GROUNDWATER FLOW REGIME

The length and width of the plume are affected by the groundwater velocity and the aquifer's hydraulic conductivity. The plume is more elongated in groundwater with high velocity than in groundwater with low velocity. The plume also tends to move slower in formations with low hydraulic conductivity than in formations with high hydraulic conductivity. A higher hydraulic conductivity can result in more rapid movement and a longer and narrower plume (Palmer 1992). The contaminant plume usually moves in the same direction as groundwater; however, this movement may not occur with a DNAPL that can sink to the bottom of the aquifer and flow by gravity in the opposite direction to groundwater flow.

Perched water is another important consideration in the effect of a groundwater flow regime on a contaminant plume. Perched water does not usually follow the regional groundwater flow direction but rather flows along an interface of hydraulic conductivity contrast. Therefore, a contaminant plume present in perched water can be moving in a different direction than the regional groundwater gradient. Groundwater fluctuations can move trapped contaminants from the vadose zone to the saturated zone.

GEOLOGY

The behavior of a contaminant plume depends largely on the type of geological profile through which it is moving. Geological structures such as dipping beds, faults, crossbedding, and facies can affect the rate and direction of a migrating plume. Dipping beds can change the direction of a migrating plume. Faults can act as a barrier or a conduit to the contaminant plume depending on the material in the fault. Interbedded clay lenses in a permeable sand formation can split or retard a sinking contaminant plume and change its shape and course. Fractures and cracks in fractured bedrock formations can act as a conduit to the contaminant plume depending on their size and interconnections. Interaquifer exchange can move a plume of contamination from formations with the greatest hydraulic head to formations of a lesser hydraulic head (Deutsche 1961).

—Ahmed Hamidi

References

Anderson, M.P. 1979. Using models to simulate the movement of contaminants through groundwater systems. *CRC Crit. Rev. Env. Control 9*, no. 2:97–156.

Bachmat, Y., and J. Bear. 1964. The general equations of hydrodynamic dispersion in the homogeneous isotropic porous mediums. *J. Geophys. Res. 69*, no. 12:2561–2567.

Bear, J. 1979. *Hydraulics of groundwater.* New York: McGraw-Hill.

CONCAWE Secretariat. 1974. *Inland oil spill clean-up manual.* Report no. 4/74. The Hague, Netherlands: CONCAWE.

Deutsche, M. 1961. Incidents of chromium contamination of groundwater in Michigan. Proceedings of Symposium on Groundwater Contamination, April. Cincinnati, Ohio: U.S. Dept. of Health, Education and Welfare.

Dragun, J. 1988. *The soil chemistry of hazardous materials.* Silver Spring, Md.: Hazardous Materials Control Research Institute.

Fenestra, S., and J.A. Cherry. 1988. Subsurface contamination by dense-non aqueous phase liquid (DNAPL) chemicals. International Groundwater Symposium, May. Halifax, Nova Scotia: International Association of Hydrogeologists.

Javandel, I., C. Doughtly, and C.F. Tsang. 1984. *Groundwater transport: Handbook of mathematical models.* Water Resources Monograph 10. Washington, D.C.: American Geophysical Union.

Montgomery, J.H., and L.M. Wekom. 1989. *Groundwater chemicals desk reference.* Chelsea, Mich.: Lewis Publishers.

Palmer, C.M. 1992. Principles of contaminant hydrogeology. Chelsea, Mich.: Lewis Publishers.

Perkins, T.K., and O.C. Johnston. 1963. A review of diffusion and dispersion in porous media. *Soc. Pet. Eng. J. 3*:70–84.

Robinson, R.A., and R.H. Stokes. 1965. *Electrolytes solutions.* 2d ed. London: Butterworth.

Schwille, F. 1988. *Dense chlorinated solvents in porous and fractured media.* Chelsea, Mich.: Lewis Publishers.

4

Groundwater Investigation and Monitoring

4.1
INITIAL SITE ASSESSMENT

The purpose of a groundwater remedial investigation is to determine the nature and extent of contamination, identify current or potential problems caused by the contamination, and assist in the evaluation and selection of the remedial action. The remedial investigation generally has two phases. The first phase, called initial assessment, involves the use of existing site information and initial field screening techniques to identify potential sources of contamination; develop a conceptual understanding of the site and contamination process; and optimize subsequent, more intrusive, field investigation. The second phase involves a detailed subsurface investigation to assess the magnitude and extent of contamination and evaluate remedial actions.

Interpretation of Existing Information

Because the potential costs involved in groundwater remedial investigations are large, the best use of existing data and information must be made. Existing information and data can be site-specific, such as records of operations and records of previous investigations, or regional including surveys of geology, hydrology, surface soils, and meteorology. Existing data, however, can vary in quality; therefore, a thorough review and interpretation of these data prior to the investigation is necessary.

SITE-SPECIFIC INFORMATION

Existing data on site history can provide useful information on potential causes and sources of groundwater contamination. Data that should be collected include old maps and aerial photographs, interviews with present and former employees at the plant site, records of operations, records of product losses and spills, waste disposal practices, and the list of contaminants generated over the operating history of the site. The inventory must also include a history of the raw materials used and wastes disposed of over the years as industrial processes changed. Particular attention should be paid to potential sources of ground-

water contamination such as locations of abandoned and active landfills and wastewater impoundments, buried product pipelines, old sewers, tanks, cesspools, dry wells, product storage areas, product loading areas, storm water collection areas, and previous spill areas.

In addition, foundation borings or construction details of supply wells can provide firsthand information on the types and characteristics of subsurface soils and groundwater at the site. Chemical data may be available from the results of previous monitoring activities at the site or at adjacent properties. These data should be analyzed and plotted on base maps and used to estimate background groundwater and soil quality.

REGIONAL INFORMATION

Regional information can be used to identify potential offsite sources of contamination and to provide background information on regional geology, hydrology, surface soils, and meteorology. This information can provide insight into the complexities of the groundwater contamination and help guide future site investigations.

A regional inventory of potential offsite sources of contamination can be developed through aerial photographs, land-use maps, and field inspections. Old aerial photographs are especially useful because they may be the only means of identifying abandoned facilities such as old landfills, lagoons, and industrial facilities. Land-use maps can identify unsewered residential areas that can be a potential source of contamination, especially where organic chemical septic tank cleaners have been used. Topographic maps can identify surface drainage patterns that can carry contaminants to the plant site and recharge the underlying groundwater system.

Regional geologic reports, maps, and cross sections can provide details on the regional subsurface geology including areal extent, thickness, composition, and structure of the geological units present in the region. Regional hydrogeologic reports can provide information on the regional groundwater flow direction and quality as well as the groundwater usage in the region. A survey of state files can reveal long-term groundwater quality problems in the general area of the plant site.

Soil maps can be used to evaluate the migration potential of contaminants through the unsaturated zone. Climatological data can be used to determine precipitation rates and patterns as well as surface runoff and groundwater recharge rates. In addition, climatological data can be used to determine evapotranspiration rates from shallow groundwater tables and their effect on the gradient and direction of groundwater.

An inventory of regional information is available from state and federal agencies such as the U.S. Geological Survey (USGS), the U.S. EPA, the U.S. Department of Agriculture (USDA), and the Soil Conservation Service (SCS). Other sources of information include computerized databases on environmental regulations and technical information on a variety of chemical compounds (Lynne et al. 1991). Examples of these databases include the Computer-Aided Environmental Legislative Data System (CELDS), which provides a collection of abstracted federal and state environmental regulations and standards; HAZARDLINE, which provides information on over 500 hazardous workplace substances as defined by the Occupational Safety and Health Administration (OSHA); and the Chemical Information System (CIS), which provides a variety of subjects related to chemistry.

Initial Field Screening

Data collection in a groundwater remedial investigation can begin with minimally intrusive techniques, called initial field screening techniques. These techniques are less expensive than the more intrusive techniques such as soil borings, test pits, and well monitoring. In addition, field screening techniques provide information which streamlines data collection and optimizes the use of intrusive techniques. The principal categories of initial field screening techniques include surface and downhole geophysical surveys and onsite chemical screening, such as a soil-gas survey.

SURFACE GEOPHYSICAL SURVEYS

Surface geophysical surveys are applied at the surface to provide a rapid reconnaissance of the hydrogeologic conditions at the site, such as depth to bedrock, degree of weathering, and the presence of clay lenses, fracture zones, or buried waste. In addition, surface geophysical surveys can be used to detect and map inorganic contaminant plumes, obtain the flow direction, and estimate the concentration gradients (Benson et al. 1985).

Surface geophysical surveys include electromagnetic conductivity, electrical resistivity, seismic refraction, and ground-penetrating radar as described in Pitchford, Mazzella, and Scarbrough (1988); Benson, Glaccum, and

Noel (1984); and the U.S. EPA desk reference guide on subsurface characterization and monitoring techniques (1993a). A description of the most commonly used surface geophysical surveys follows.

Electromagnetic (EM) Methods

The EM methods use a transmitter coil to generate an electromagnetic field that induces eddy currents in the ground below the instrument. A receiver coil measures secondary electromagnetic fields created by the eddy currents and produces an output voltage that can be related to variations in subsurface conductivity as shown in Figure 4.1.1. Variations in subsurface conductivity may be caused by changes in the basic soil or rock types, thickness of the soil and rock layers, moisture content, fluid conductivity, and depth to the water table.

Environmental engineers can use EM surveys to obtain data by profiling or sounding. In profiling, the engineer makes measurements at a number of stations along a survey line to map lateral changes in the subsurface electrical conductivity to a given depth. In sounding mode, the engineer places the instrument at one location and takes measurements at increasing depths, by changing coil orientation or coil spacing, to map vertical changes in electrical conductivity and, therefore, the soil and rock type at that location.

An advantage of the EM methods is that the surveys can be done quickly because direct contact of the instrument with the ground is not required. The disadvantage, however, is that the EM surveys are susceptible to the presence of metals and powerlines on the surface of the ground.

Electrical Resistivity (ER) Methods

In ER methods, environmental engineers measure the resistivity of subsurface materials by injecting an electrical current into the ground through a pair of surface electrodes (current electrodes) and measuring the resulting potential field (voltage) from a second pair of electrodes (potential electrodes) as shown in Figure 4.1.2. Several types of electrode geometries can be used for resistivity measurements including the Wenner, Schlumberger, dipole, and others. The Wenner array is the simplest in terms of geometry and consists of four electrodes spaced equally in a line.

The ER measurements are a function of the soil or rock types, thickness of the soil and rock layers, moisture content, fluid conductivity, and depth to the water table. The ER of a geological formation is calculated based on the electrode separation, the geometry of the electrode array, the applied current, and the measured voltage.

As with the EM surveys, environmental engineers can use the ER surveys to obtain data by profiling or sounding. In profiling, engineers take measurements at a num-

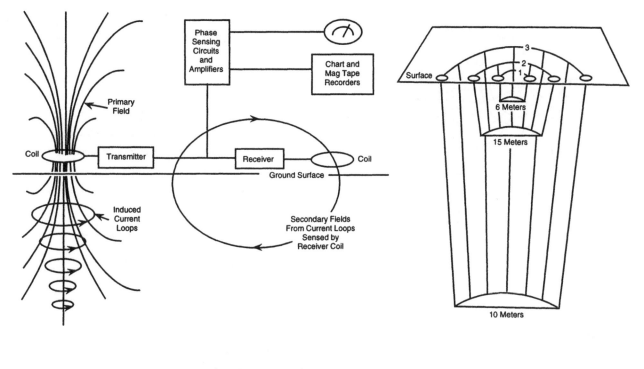

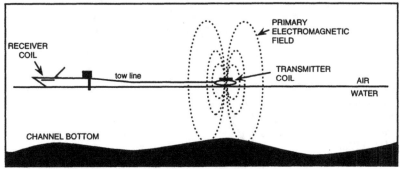

FIG. 4.1.1 Electromagnetic survey. (Reprinted from U.S. Environmental Protection Agency, 1993, *Subsurface characterization and monitoring techniques, a desk reference guide*, U.S. EPA/625/R-93/003a [May] U.S. EPA.)

ber of stations along a survey line to map lateral changes in the subsurface electrical properties to a given depth. Then, they can use the data to delineate hydrogeological anomalies or map inorganic plumes. Sounding measurements, on the other hand, are made at increasing depths so that engineers can map vertical changes in electrical properties. Engineers use data from sounding measurements to determine the depth, thickness, and type of soil or rock layer at the site. The data from ER surveys can be interpreted with the use of computer models or master curves to create geoelectric sections (Orellana and Mooney 1966). These sections illustrate changes in the vertical and lateral resistivity conditions at the site.

The ER surveys are useful for identifying shallow contaminated groundwater bodies where (1) a significant contrast exists in water quality; (2) the water table is less than 40 feet deep; (3) the geology of the water table aquifer is relatively homogeneous; and (4) local interferences, such as buried pipelines, power lines, or metal fences, are not present.

The advantages of the ER methods are that they are well established and their equipment is inexpensive, mobile, and easy to operate and provides relatively rapid areal coverage. In addition, the ER methods are superior to the EM methods for detecting thin resistive layers. The disadvantage, however, is that continuous profiling is not possible, and the requirement for ground contact can cause problems in resistive material and generally makes the ER surveys slower to use than the EM surveys. Furthermore, use of the ER methods is limited in wet weather and on paved areas, and the methods are less sensitive to conductive pollutants than the EM methods.

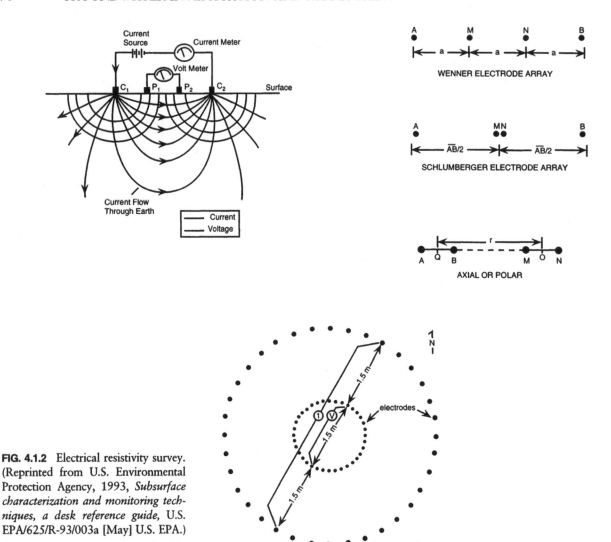

FIG. 4.1.2 Electrical resistivity survey. (Reprinted from U.S. Environmental Protection Agency, 1993, *Subsurface characterization and monitoring techniques, a desk reference guide*, U.S. EPA/625/R-93/003a [May] U.S. EPA.)

Seismic Refraction (SR) Methods

Environmental engineers often use the SR methods to determine the top of bedrock or depth of the water table, locate fractures or faults, and characterize the type of rock or degree of weathering. The SR methods are based on the fact that elastic waves travel through different earth materials at different velocities; the denser the material, the higher the wave velocity.

The elastic waves are initiated by an energy source (hammer or controlled explosive charge) at the ground surface. A set of receivers, called geophones, is set up in a line radiating outward from the energy source as shown in Figure 4.1.3. Waves initiated at the surface and refracted at the critical angle by a high-velocity layer at a depth reach the more distant geophones quicker than the waves that travel directly through the low-velocity surface layer. The time between the shock and the arrival of the elastic wave at a geophone is recorded on a seismograph. Using a set of seismograph records, engineers can derive a graph of arrival time versus distance from the shot point to the geophone. They can then analyze the line segments, slope, and break points in the graph to identify the number of layers and the depth of each layer. In addition, they can use typical seismic velocity ranges to determine the type of soil of each layer (U.S. EPA 1993a).

The advantages of SR methods are that the equipment is readily available, portable, and relatively inexpensive. In addition, the methods are accurate and provide rapid areal coverage with depths of penetration up to 30 meters. The disadvantage, however, is that the resolution might be obscured by layer sequences where the velocity of the layers decreases with depth, and thin layers, called blind zones, might not be detected. Furthermore, the methods are susceptible to noise from adjacent areas (such as construction activities) and do not detect contaminants in groundwater.

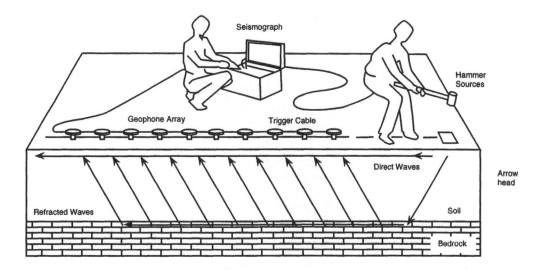

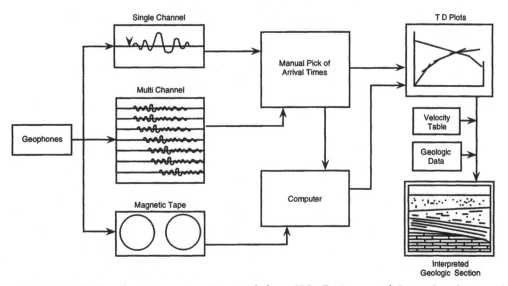

FIG. 4.1.3 Seismic refraction survey. (Reprinted from U.S. Environmental Protection Agency, 1993, *Subsurface characterization and monitoring techniques, a desk reference guide,* U.S. EPA/625/R-93/003a [May] U.S. EPA.)

Ground Penetrating Radar (GPR) Methods

Environmental engineers often use the GPR methods to locate buried objects, map the depth to shallow water tables, and delineate soil horizons. The principles involved in GPR technology are similar to those in seismic refraction, except that in GPR, electromagnetic energy is used instead of acoustic energy, and the resulting image is relatively easy to interpret.

In a GPR survey, a transmitting and a receiving antenna are dragged along the ground surface as shown in Figure 4.1.4. The small transmitting antenna radiates short pulses of high-frequency radio waves into the ground, and the receiving antenna records variations in the reflected return signal. The attenuation loss of the signal in the ground increases with ground conductivity and with frequency for a given material. Changes in ground electric conductivity are associated with natural hydrogeological conditions such as bedding cementation, moisture, clay content, voids, and fractures. Therefore, an interface between two soil or rock layers with sufficient contrast in electric conductivity shows up in the radar profile (Benson and Glaccum 1979).

The advantages of the GPR methods include rapid areal coverage, where site conditions are favorable, and great

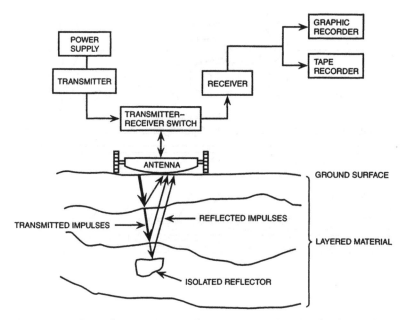

FIG. 4.1.4 Ground-penetrating radar survey. (Reprinted from U.S. Environmental Protection Agency, 1993, *Subsurface characterization and monitoring techniques, a desk reference guide,* U.S. EPA/625/R-93/003a [May] U.S. EPA.)

resolution and penetration in dry, sandy, or rocky areas. The use of GPR, however, is limited in moist and clayey soils and soils with high electrical conductivity.

DOWNHOLE GEOPHYSICAL SURVEYS

Downhole geophysical surveys provide localized details on soil, rock, or fluid along the length of an existing monitoring well or a borehole. The surveys can also identify permeable zones, such as sand lenses in glacial tills, weathered zones, and fractures or solution cavities in rocks.

Several downhole logging techniques are available including nuclear, electromagnetic, and acoustic or seismic as described in Keys and MacCary (1976) and the U.S. EPA desk reference guide on subsurface characterization and monitoring techniques (1993a). Some of these techniques provide measurements from inside plastic or steel casing, and some allow measurements in the unsaturated zone as well as the saturated zone. A description of the most commonly used logs follows.

Nuclear Logging Methods

Nuclear logging includes methods that detect the presence of unstable isotopes or create such isotopes in the vicinity of a borehole. Several nuclear logging techniques are available including natural gamma logs, gamma–gamma logs, and neutron–neutron logs. Natural gamma logs are probably the most common nuclear methods used in groundwater studies.

Environmental engineers use natural gamma logging, in general, to identify lithology and stratigraphic correlation and, in particular, to evaluate the presence, variability, and integrity of clays and shales.

The natural gamma log records the amount of natural gamma radiation emitted by rocks and unconsolidated materials from a borehole. Different formations can be distinguished from different levels of natural radioactivity as shown in Figure 4.1.5. The gamma-emitting radioisotopes normally found in all rocks and unconsolidated materials are potassium-40 and daughter products of the uranium and thorium decay series (Benson 1991). Clays and shales concentrate these heavy radioactive elements through the process of ion exchange and adsorption; therefore, their natural gamma activity is much higher than that of other materials.

The natural gamma log instrumentation is relatively simple and inexpensive and involves radiation detection only. However, only qualitative analysis is possible with this method, and the sensitivity of the probe is reduced by large diameter holes, drilling fluid, and casing (U.S. EPA 1993a).

Electromagnetic Logging Methods

As with the EM method, the electromagnetic logging method measures the electrical conductivity of soil or rock in open or polyvinyl-chloride- (PVC) cased boreholes above or below the water table. Environmental engineers use this method to perform lithological characterization, locate the zones of saturation, and perform chemical char-

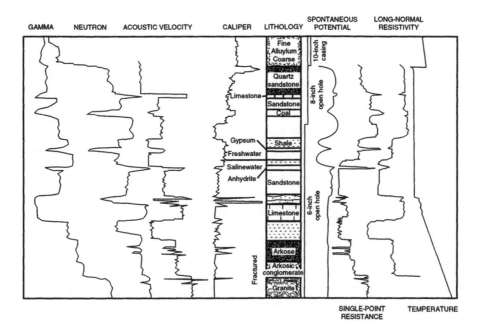

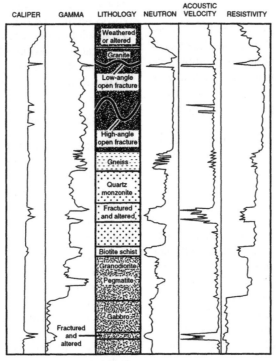

FIG. 4.1.5 Well log suites in sedimentary and fractured rocks. (Reprinted from U.S. Environmental Protection Agency, 1993, *Subsurface characterization and monitoring techniques, a desk reference guide*, U.S. EPA/625/R-93/003a [May] U.S. EPA.)

acterization of groundwater. Several electromagnetic logging techniques are available including induction logs, microwave-sensing logs, nuclear magnetic resonance logs, and surface-borehole logs. Induction logs are probably the most common electromagnetic methods used in groundwater studies.

The probe in an induction log contains a transmitter coil on the upper part, which induces eddy current in the formation around the borehole, and a receiver on the lower part. Engineers measure conductivity using the same principles as the EM methods. Because the response of the log is a function of the specific conductance of the pore fluids, it is an excellent indicator of the presence of inorganic contaminants (Benson 1991). Variations in conductivity with depth also indicate changes in clay content, permeability of a formation, or fractures.

Other Logging Methods

Several other types of logging techniques are useful for characterizing lithology and hydrogeology inside a well or a borehole. Examples of these logs include caliper logs, temperature logs, fluid-flow logs, and borehole television logs.

A caliper log provides information on the diameter, lithology, fractures, and construction details of an open borehole. Many types of caliper logs are available including mechanical, electric, and acoustic. The mechanical caliper is the most commonly used. The probe in a mechanical caliper consists of spring-loaded arms which extend from the logging tool so that they follow the sides of the borehole. Mechanical caliper tools have from one to six arms and can measure variations as small as $\frac{1}{4}$ inch in borehole diameter.

A temperature log can provide a continuous record of the temperature of the fluid inside the borehole or well. Environmental engineers can use changes in temperature to identify leaks in the casing where damage or corrosion has occurred.

Fluid-flow logs measure the fluid flow within a borehole or well (Keys and MacCary 1976). Examples of such logs include thermal and electromagnetic borehole flowmeters that sense water movement either vertically or horizontally (or both) at low velocities. Fluid-flow measurements can locate zones of high permeability (fractures) and areas of leakage in artisan wells.

Borehole television camera logs allow visual inspection of a borehole or well for fractures or casing defects (Morahan and Dorrier 1984).

ONSITE CHEMICAL SURVEYS

Environmental engineers are increasingly using onsite chemical surveys as field screening techniques to pinpoint source areas or approximately delineate the extent of existing contaminant plumes. The use of onsite chemical surveys optimizes the number of samples taken by more expensive intrusive techniques and sent to the laboratory for confirmatory chemical analysis. Several techniques are available for volatile and nonvolatile organics as well as for inorganic compounds.

Onsite chemical screening techniques vary from qualitative chemical analyses using indicators such as organic vapor analyzers (OVAs) or HNU meters to more quantitative soil-gas surveys using gas chromatography and mass spectrometry (GC/MS).

Qualitative Onsite Chemical Surveys

Generally, environmental engineers use these field screening techniques to collect preliminary site information and guide future and more intrusive field investigations. Engineers can measure the pH of the soil, waste, or groundwater in the field with a pH meter and use the results of these measurements to characterize the subsurface environment or classify the corrosivity of waste materials. They can also electrometrically measure the Eh of groundwater in the field using a platinum electrode and a reference electrode (Holm, George, and Barcelona 1986; Ritchey 1986). Then, they can use the results of the measurements to characterize oxidation-reduction conditions in the subsurface and evaluate the potential for mobility of heavy metals in groundwater.

OVAs, photoionization detectors (PID/HNU meter), flame ionization detectors (FIDs/OVAs), argon ionization detectors (AIDs), and combustible gas indicators (EDs) are all total organic vapor survey instruments that locate source areas of volatile compounds within the vadose zone or track these compounds within groundwater (U.S. EPA 1993b).

Test kits are commercially available for preliminary field screening of many inorganic compounds (Hatch kits) and some organic compounds (Handy kits). These kits are based on the principles of colorimetry. Colorimetry involves mixing the reagents of known concentrations with a test solution in specified amounts. This mixing results in chemical reactions in which the color of the solution is a function of the concentration of the analyte of interest (Davis et al. 1985; Fishman and Friedman 1989).

Soil-Gas Surveys

Environmental engineers use soil-gas surveys to locate source areas of volatile compounds within the vadose zone, track plumes of volatile compounds in groundwater, identify migration patterns of landfill gases, and optimize the number and location of more expensive and intrusive monitoring points such as soil borings and groundwater monitoring wells.

Soil-gas surveys are based on several in situ soil sampling techniques such as headspace analysis, surface flux chambers, downhole flux chambers, surface accumulators, and suction ground probes. The most commonly used techniques, however, are the surface accumulators and the suction probes.

Surface accumulators involve the passive sampling of soil gas by trapping volatile organic compounds (VOCs) onto an adsorbent contained within an inverted glass tube (Zdeb 1987). The inverted glass tube is buried in the soil for a few days to weeks. The adsorbent consists of a ferromagnetic wire coated with activated charcoal and is contained in an inverted test tube. The adsorbent passively collects diffusing VOCs which adsorb onto the activated charcoal. After a few days or weeks, the glass tube is sealed and taken to the laboratory for VOC analysis.

Ground probe sampling techniques for soil gas involve inserting a tube into the ground and pumping the soil gas with a vacuum pump. Engineers then analyze the extracted gas in the field for VOCs using portable analytical instruments. The probes can be manually or pneumatically driven or installed in boreholes. Grab samples can be taken at the same depth (or at different depths) at several locations for areal (or vertical) characterization of soil-gas concentrations.

The vertical and horizontal spacing of the probes can be affected by many factors such as soil moisture and organic matter content, presence of perched water, depth to groundwater, permeability of the subsurface materials, and the Henry's Law constant of the VOC in question (Silka 1986). The upward diffusion of vapors is usually blocked by soil strata containing a finer grained soil with a higher moisture content or higher organic carbon content.

—Ahmed Hamidi

References

Benson, R.C. 1991. Remote sensing and geophysical methods for evaluation of subsurface conditions. In *Practical handbook of groundwater monitoring*, edited by David Nielson. Chelsea, Mich.: Lewis Publishers.

Benson, R.C., and R.A. Glaccum. 1979. Radar surveys for geotechnical site assessment. Proceedings of the Geophysical Methods in Geotechnical Engineering, Specialty Session, 161–178. Atlanta, Ga.: American Society of Civil Engineers.

Benson, R.C., R.A. Glaccum, and M.R. Noel. 1984. *Geophysical techniques for sensing buried wastes and waste migration*. Worthington, Ohio: National Water Well Association.

Benson, R.C. et al. 1985. Correlation between geophysical measurements and laboratory water sample analysis. Proceedings of the National Water Well Association, Environment Protection Agency Conference on Surface and Borehole Geophysical Methods in Groundwater Investigation. National Water Well Association.

Davis, S.N. et al. 1985. *Introduction to groundwater tracers*. EPA/600/2-85/022, NTIS PB86-100591.

Fishman, M.J., and L.C. Friedman, eds. 1989. *Methods for determination of inorganic substances in water and fluvial sediments*. 3d ed. U.S. Geological Survey Techniques of Water Resources Investigations, TWRI 5-A1.

Holm, T.R., G.K. George, and M.J. Barcelona. 1986. *Dissolved oxygen and oxidation-reduction potentials in groundwater*. EPA/600/2-86/042, NTIS PB86-179678.

Keys, W.S., and L.M. MacCary. 1976. *Application of borehole geophysics to water resources investigations*. Techniques of Water Resources Investigations of the United States Geophysical Survey.

Lynne, M. et al. 1991. The overall philosophy and purpose of site investigation. In *Practical handbook of groundwater monitoring*, edited by David Nielsen. Chelsea, Mich.: Lewis Publishers.

Morahan, T., and R.C. Dorrier. 1984. The application of television borehole logging to groundwater monitoring programs. *Groundwater Monitoring Review* 4, no. 4:172–175.

Orellana, E., and H.M. Mooney. 1966. *Master tables and curves for vertical electrical sounding over layered structures*. Madrid, Spain: Interciencia.

Pitchford, A.M., A.T. Mazzella, and K.R. Scarbrough. 1988. *Soil and geophysical techniques for detection of subsurface organic contamination*. USEPA/600/4-88-019, NTIS. U.S. EPA.

Ritchey, J.D. 1986. Electronic sensing devices used for in situ groundwater monitoring. *Groundwater Monitoring Review* 6, no. 2:108–113.

Silka, L.R. 1986. Simulation of the movement of volatile organic vapor through the unsaturated zone as it pertains to soil–gas surveys. Proceedings of the NWWA/API Conference on Petroleum Hydrocarbons and Organic Chemicals in Groundwater: Prevention, Detection, and Restoration. Dublin, Ohio: National Water Well Association.

U.S. Environmental Protection Agency. 1993a. *Subsurface characterization and monitoring techniques, a desk reference guide*. USEPA/625/R-93/003a (May). U.S. EPA.

———. 1993b. *Subsurface characterization and monitoring techniques, a desk reference guide*, Vol. 2. USEPA/625/R-93/003b (May). U.S. EPA.

Zdeb, T.F. 1987. Multi-depth soil–gas analysis using passive and dynamic sampling techniques. Proceedings of Petroleum Hydrocarbons and Organic Chemicals in Groundwater: Prevention, Detection, and Restoration. Dublin, Ohio: National Water Well Association.

4.2
SUBSURFACE SITE INVESTIGATION

The purpose of a subsurface investigation is to collect samples and obtain actual quantitative measurements of chemical concentrations, hydraulic parameters, and lithological data within a particular hydrogeologic strata or group of strata. Environmental engineers can use these samples and measurements to assess the magnitude and extent of groundwater or soil contamination and support the selection and design of engineering options for remediation.

Engineers can conduct subsurface investigations using temporary groundwater and soil sampling techniques such as HydroPunch, soil probes, and cone penetrometers (Edge and Cordy 1989) or more permanent techniques such as the installation of monitoring wells and soil borings. Temporary techniques are less expensive but less reliable; therefore, they are usually used for screening purposes and the optimization of the location and number of permanent systems. Permanent techniques, on the other hand, are more expensive and more reliable; therefore, their use is usually limited to confirm actual concentrations and subsurface conditions.

Subsurface investigations involve several field activities such as drilling, installation, development, and sampling of monitoring wells. These activities are intrusive to the subsurface environment; therefore, engineers should conduct them with care to prevent cross-contamination and obtain representative groundwater and soil samples that retain both the physical and chemical properties of the subsurface environment. A description of these field activities follows.

Subsurface Drilling

Subsurface drilling for groundwater remedial investigations uses much of the same technology as conventional geotechnical exploration but with some significant differences. Geotechnical exploration requires the collection of an intact physical specimen which can be tested for geotechnical properties. In comparison, groundwater remedial investigations require that the specimen also be representative of existing conditions and valid for chemical analysis. Therefore, the selection of drilling methods and sampling protocols in a groundwater remedial investigation is more restrictive and should be based upon site-specific conditions and the type of testing to be done.

The criteria used in the selection of a drilling method include the type of geological formation, depth of drilling, depth of screen setting, types of pollutants expected, accessibility to the site, and availability of drilling equipment. The following section briefly describes the drilling methods used in groundwater remedial investigations.

DRILLING METHODS

Several drilling methods are used in groundwater remedial investigations including air rotary, direct mud-rotary, reverse mud-rotary, hollow-stem augers, solid-stem augers, and cable tools among others (Davis, Jehn, and Smith 1991). The following discussion focuses on the two methods most commonly used for monitoring well installations: hollow-stem auger and direct mud-rotary.

Hollow-Stem Auger

The hollow-stem auger is a form of continuous flight auger usually used for drilling monitoring wells in unconsolidated materials. The auger consists of a tubular steel center shaft or axle around which is welded a continuous steel strip in the form of a helix, also known as flight, as shown in Figure 4.2.1. As the auger column rotates and axially advances in the ground, the dug material is simultaneously conveyed to the surface by the helix.

The main advantage of hollow-stem auger drilling is that no drilling fluids or lubricants are used; therefore, no contaminants are introduced into the aquifer. In addition, the hollow stem of the auger allows sampling of soil material as the borehole is advanced and installation of casings and screens for monitoring wells when the required depth has been reached. The drill head, or cutting bit, lo-

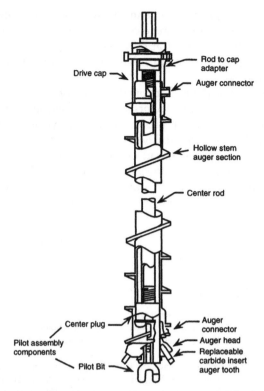

FIG. 4.2.1 Hollow-stem auger system. (Reprinted from U.S. Environmental Protection Agency, 1993, *Subsurface characterization and monitoring techniques, a desk reference guide,* Vol. 1, USEPA/625/R-93/003a [May] U.S. EPA.)

cated at the bottom of the auger can be removed (tripped) through the center of the auger to the surface. This feature allows the auger to stay in place providing an open, cased hole into which samplers, downhole drive hammers, casings, screens, and other instruments can be inserted.

The hollow-stem auger cannot be used, however, in consolidated, rock, or well-cemented formations. In addition, depths are usually limited to no more than 150 feet, and vertical leakage of water through the borehole during drilling is likely to occur.

Direct Mud-Rotary

Direct mud-rotary drilling is a drilling method in which a fluid is forced down the drill stem, out through the bit, and back up the borehole to remove the cuttings as shown in Figure 4.2.2. The cuttings are removed by settling in a sedimentation tank or pond, and the fluid is circulated back down the drill stem. The drilling fluid can be a liquid, such water or mud (water with special additives, e.g., bentonite and polymers), or it can be gas, such as air or foam (air with additives, e.g., detergents) (Davis, Jehn, and Smith 1991).

Mud-rotary drilling is a flexible and rapid drilling method in all types of geologic materials and depth ranges. The circulating fluid serves to cool and lubricate the bit,

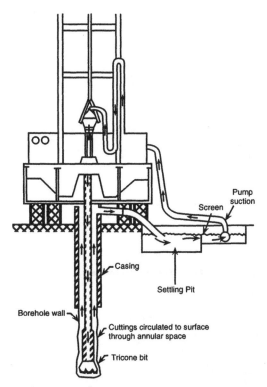

FIG. 4.2.2 Direct mud-rotary circulation system. (Reprinted from U.S. Environmental Protection Agency, 1993, *Subsurface characterization and monitoring techniques, a desk reference guide*, Vol. 1, USEPA/625/R-93/003a [May] U.S. EPA.)

stabilize the borehole, remove the cuttings, and prevent the inflow of formation fluids, thus minimizing cross-contamination of aquifers. In addition, samples can be obtained directly from the circulated fluid when a sample-collecting device is placed in the discharge pipe before the settling tank.

Mud-rotary drilling, however, requires the introduction of some foreign liquids into the aquifer, which can compromise the validity of subsequent monitoring well samples. In addition, contaminants might be circulated with the fluid, and the collection of representative samples is difficult due to the mixing of drill cuttings. Other limitations of mud-rotary drilling include the inability to provide information on the position of the water table and the loss of drilling fluids in fractured materials.

SOIL SAMPLING

Soil samples are usually taken at regular intervals during drilling to be analyzed for chemical composition and tested for physical properties such as particle size distribution, textural classification, and hydraulic conductivity. The samples are generally taken from the bottom of the borehole and at the necessary depth when the sampling device is driven with the aid of a 140-pound hammer. The hammer is connected to the sampling device by drill rods. The

number of hammer blows, usually counted for each 6-inch increment of the total drive, indicates the compaction and density of the formation being penetrated.

The most commonly used soil sampling devices are the split-spoon sampler and the Shelby tube. The split-spoon sampler is a 12- or 18-inch long hollow cylinder consisting of two equal semicylindrical halves held together at each end with threaded couplings. The sampler is lowered to the bottom of the borehole and driven with a hammer to the necessary depth. When the sampler is brought to the surface, it is disassembled, or split, to remove the soil sample. Split-spoon sampling provides representative soil samples for physical or chemical testing. The samples, however, are disturbed; therefore, the results of the analysis should be used with caution. When the same sampler is used to collect different samples, the engineer should decontaminate it after each sampling event to prevent cross-contamination.

The Shelby tube sampler is a thin-walled tube made of steel, aluminum, brass, or stainless steel. The cutting edge of the tube is sharpened, and the upper end is attached to a coupling head by cap screws. The sampler must meet certain criteria, such as a clearance ratio of 0.5 to 1.5 and end area ratio of 10, to ensure the least disturbance to the sample (U.S. EPA 1993). The sample collection procedure is similar to split-spoon sampling except that the tube is pushed into the soil by the weight of the drill rig rather than driven. When the sampler is brought to the surface, the sample is sealed and preserved for laboratory analysis. Shelby tube sampling is used for soil analyses that require undisturbed soil samples, such as hydraulic conductivity testing.

Monitoring Well Installation

The installation of groundwater monitoring wells involves selecting the location and number of wells, drilling the boreholes, installing the casings and screens, placing the filter pack materials and annular seals, and finally developing the wells. Each of these steps should be designed to meet the objectives of the monitoring program and suit the conditions of the site. A typical monitoring well design is shown in Figure 4.2.3.

WELL LOCATION AND NUMBER

The selection of the proper number and locations of monitoring wells is obviously one of the most important decisions in any groundwater monitoring program or groundwater remedial investigation. For monitoring purposes, most guidances suggest a minimum of four monitoring wells per potential source of groundwater contamination (U.S. EPA 1986). Three of these wells are placed downgradient of the potential source, and one is placed upgradient. The purpose of these wells is to provide information on the background quality of the groundwater

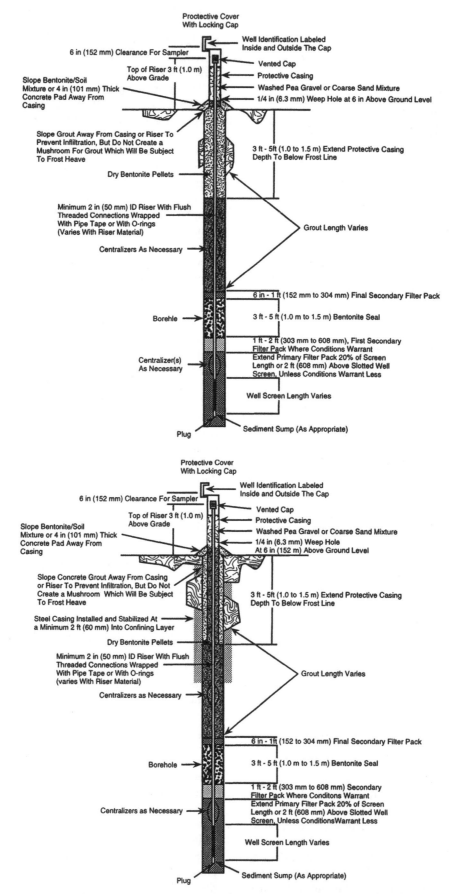

FIG. 4.2.3 Typical monitoring well design and construction detail. (Reprinted from U.S. Environmental Protection Agency, 1993, *Subsurface characterization and monitoring techniques, a desk reference guide,* Vol. 1, USEPA/625/R-93/003a [May] U.S. EPA.)

upgradient of the source and detect or monitor any contaminant plumes emanating from the source.

In a remedial investigation, however, the preliminary selection of the location and number of wells needed to delineate and monitor the plume is usually based on the results of initial field screening techniques such as gas surveys, HydroPunch, geophysical surveys, and borings. Environmental engineers use the data from these investigations to estimate the extent of the contaminant plume and establish the basic hydrogeologic parameters of the site. Once the hydrogeology of the site is understood and the migration path of the suspected contaminant plume is established, the location and number of wells can be finalized. In general, the more complicated the hydrogeology, the more complex the migration path of the contaminant plume and the greater the number of required monitoring wells (Barcelona et al. 1990).

CASINGS AND SCREENS

The purpose of the casing and screen in a groundwater monitoring well is to provide access from the surface to the groundwater in order to collect groundwater samples or measure groundwater elevations. The casing prevents geologic materials from collapsing into the borehole, while the screen allows groundwater to enter the monitoring well. The screen is generally attached to the subsurface end of the casing.

Several types of casing and screen materials are available including stainless steel, galvanized steel, carbon steel, PVC, Teflon, and aluminum. Selecting the monitoring well casing and screen material depends on the physical strength and chemical reactivity of the material under subsurface conditions. With regard to physical strength, the casing and screen should be able to withstand the forces exerted on them by the surrounding geologic materials. These forces can be significant for deep monitoring wells (greater than 30 meters). Nielson and Schalla (1991) provide data on the physical strength of different types of casing and screen materials.

With regard to chemical reactivity, the material of the casing and screen should neither adsorb nor leach chemical constituents which would bias the representativeness of the samples collected. In addition, the material must be durable enough to endure chemical attacks (corrosion or chemical degradation) from the natural chemical constituents or the contaminants in the groundwater. Teflon is probably the most chemically resistant material used in monitoring well installation, but the cost of Teflon is high (Barcelona et al. 1990). Stainless steel offers good strength and chemical resistance in most environments (except in highly acidic conditions), but it too is expensive. Galvanized steel is less expensive; however, it can impart iron, manganese, zinc, and cadmium to many waters. PVC has good chemical resistance except to low molecular weight

ketones, aldehydes, and chlorinated solvents (Miller 1982).

Two types of screens are commonly used in monitoring wells: machine-slotted pipes and continuous-clot wire-wound screens. Machine-slotted pipes are readily available and inexpensive, but the low amount of open area in these screens makes development of the well difficult. Continuous-slot screens, in contrast, are more efficient, but their cost is relatively high. The design of the slot size of the screen must be based on the characteristics of the filter pack material and the grain size of the stratum. The optimum slot size should provide maximum open area for water to flow through and minimum entry of fine particles into the well during piping (Nielson and Schalla 1991; Aller et al. 1991).

The depth of placement of the screen as well as its length are usually determined based on the depth and thickness of the water-bearing zone to be monitored. When the objective of the well is to monitor a potable water supply aquifer, then a longer screen, perhaps over the entire thickness of the aquifer, might be selected. On the other hand, when the objective of the well is to vertically delineate a plume, such as with cluster wells, then shorter screens at specific intervals might be selected.

The screen should be fully submerged to prevent contact between the contaminated groundwater and the atmosphere, particularly for volatile compounds. The screen is, however, extended above the water table for wells constructed to monitor floating products. In this case, the screen length and position must accommodate variations in water table elevation.

The casings are produced in various diameters (2, 4, 6, and 8 inches) and various lengths (5, 10, and 20 feet) that are joined by various coupling methods during installation. The casing diameter depends on the future use of the well, the type of pumping equipment, and the method of drilling. Small diameters (2 and 4 inches) are used for monitoring wells, while large diameters (6 and 8 inches) are used for recovery wells.

The casing must extend from the top of the screen to the ground surface level. The casing is protected at the surface by a metal protective casing or a manhole. Multiple casings are installed for wells penetrating more than one water-bearing formation. The purpose of multiple casings is to prevent a hydraulic connection and potential cross contamination between the water-bearing formations along the annular space produced by the installation of well casings. Figure 4.2.3 shows an example of a double-cased well installation where the outer casing is anchored into the confining layer before the borehole is advanced and the well is installed through the inner casing.

FILTER PACKS AND ANNULAR SEALS

Filter packs placed around the well screen allow groundwater to flow freely into the well while keeping fine particles from entering the well. Two types of filter packs are used in monitoring wells: naturally developed filter packs and artificially introduced filter packs. Naturally developed filter packs are produced in situ when the fine-grained materials around the screen are removed during the well development process. Environmental engineers construct artificial filter packs by backfilling the annular space surrounding the screen with a granular, relatively inert material such as clean silica sand.

In an artificially filter-packed well, the filter material can be selected for optimum efficiency of well operation, but the procedure of introducing the filter pack is time consuming and expensive. Furthermore, bridging can prevent complete filling around the well screen, and the filter pack material can introduce contaminants into the aquifer; a leaching test can determine whether this contamination is a problem. Naturally developed filter packs are, on the other hand, simpler, less expensive, and do not introduce new contaminants into the aquifer. However, well development for these filter packs is more difficult, and success is less assured.

Engineers can use a tremie pipe or a reverse circulation method to place the artificial filter pack. The tremie pipe method allows funneling of the material directly into the interval around the well screen. In a reverse circulation method, a mixture of sand and water is fed into the annulus around the screen, and the water entering the screen is pumped up to the surface through the casing. The engineer progressively pulls back the temporary casing (for hollow-stem augers) to expose the screen as the filter pack material builds up around the well screen.

Artificially introduced filter packs usually extend from the bottom of the screen to at least 3 to 5 feet above the bottom of the screen. This extension accounts for settlement of the filter pack material and allows a sufficient buffer zone between the well screen and the annular seal above.

After the filter pack is placed around the well screen, the engineer seals the annular space between the well casing and the formation to prevent upward or downward movement of water and contaminants along this pathway. In addition, the engineer places a surface seal of concrete around the protective casing to prevent surface drainage into the borehole. The annular seal is usually composed of bentonite or neat cement (Williams and Evans 1987). Bentonite is readily available and inexpensive but can cause constituent interference due to ion exchange. Neat cement is also readily available and inexpensive, but channeling between the casing and seal can develop due to temperature changes during the curing process (U.S. EPA 1993).

The engineer places the sealing mixture in the annular space using a side-discharge tremie pipe through which the grout is pumped from the surface. Complete sealing of the annular space is necessary to avoid potential bridging of the grout with formation material (Campbell and Lehr 1975).

WELL DEVELOPMENT

The purpose of well development is to remove the residues of drilling fluids and fine particles of filter packs so that subsequent sampling is representative of the groundwater. The development should be performed as soon as possible after the well is installed and the annular seal is cured.

Development methods include bailing, overpumping, air surging, and high-velocity jetting. In bailing, a bailer is dropped and retrieved in and out of the well causing an outward surge of water through the well screen and filter pack. Such surging forces the loosely bound fine particles through the screen and into the well where they can be removed by the bailer. Bailing has the advantage of being a simple technique which does not introduce new fluids into the aquifer. However, bailing is time consuming and ineffective in unproductive wells.

In overpumping, a submersible pump is lowered into the well and alternatively turned on and off, usually at a slightly higher rate than what the formation can deliver. This action, along with the repeated raising and lowering of the pump into the well, causes the water to move back and forth through the well screen, moving fine particles and drilling fluids into the well where they can be removed. Overpumping is convenient for small wells or poor aquifers; however, excessive pumping rates can cause well collapse, especially in deep wells.

Air surging consists of injecting compressed air in the well, causing the water column to lift almost to the surface, and shutting off the air supply to allow the column to fall back into the well. Repeated use of this technique causes an outward surging action in the well intake which forces the loosely bound fine particles through the screen and into the well where they can be removed. Environmental engineers must filter the injected air so that contaminants, such as lubricants of the compressor, are not introduced into the well.

High-velocity jetting uses nozzle devices to force a horizontal stream of water against the well screen opening. Engineers can remove the material that enters the screen in the backwash of the jet stream by pumping or bailing. High-velocity jetting is effective in removing the mud cake and breaking the bridges in the filter pack. However, this technique can introduce potential air and water contaminants to the aquifer.

Groundwater Sampling

The objective of any groundwater sampling program is to collect and analyze samples that are representative of existing groundwater conditions at the site. This goal is achieved with a sampling plan that incorporates sampling procedures designed to minimize sources of error or misrepresentation in each stage of the sampling process. The key stages of sampling involve well purging, sample collection and pretreatment, sample handling and preservation, and analysis and reporting of analytical data. A brief description of these sampling stages follows. Figure 4.2.4 presents a generalized flow diagram of groundwater sampling protocol.

PURGING

Before environmental engineers can sample a monitoring well, they must remove the water standing in the well to allow fresh water from the aquifer to enter the well. Purging is necessary because the stagnant water in the well is subject to chemical reactions from contact with well construction materials and the atmosphere for extended periods of time (Seanor and Brannaka 1983; Wilson and Dworkin 1984). The volume of water which should be removed from the well is based on the hydraulic characteristics of individual wells and geological settings (Gibb, Schuller, and Griffin 1981). A general rule is to remove three to five well volumes or to remove water until the water quality indicators, such as pH, conductance, and temperature are stable.

When purging a well, engineers should not allow the water level to drop below the level of the well screen to avoid aeration and loss of volatile or redox-sensitive compounds. In addition, the pumping rate should not exceed levels that might cause turbulent flow in the well and subsequent pressure changes and loss of dissolved gases (Meridith and Brice 1992). Overpumping can also dilute the sample or increase its turbidity because of the fine particles that may be drawn into the well.

Engineers should use the same equipment for purging and sampling to minimize the number of items that enter the well and therefore, the possibility of cross contamination. Furthermore, placing the purging device at the top of the well screen or at the top of the column of water ensures that all stagnant water is removed (Unwin and Huis 1983).

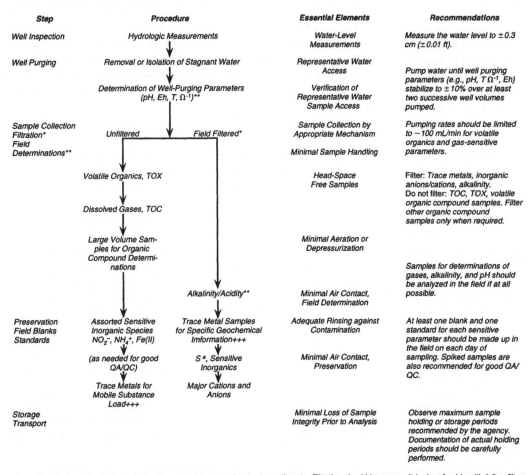

* Denotes samples that should be filtered to determine dissolved constituents. Filtration should be accomplished preferably with inline filters and pump pressure or by N_2 pressure methods. Samples for dissolved gases or volatile organics should not be filtered. In instances where well development procedures do not allow for turbidity-free samples and may bias analytical results, split samples should be spiked with standards before filtration. Both spiked samples and regular samples should be analyzed to determine recoveries from both types of handling.
** Denotes analytical determinations that should be made in the field.
+++ See Puls and Barcelona (1989).

FIG. 4.2.4 Generalized flow diagram of groundwater sampling protocol. (Reprinted from U.S. Environmental Protection Agency, 1993, *Subsurface characterization and monitoring techniques, a desk reference guide,* Vol. 1, USEPA/625/R-93/003a [May] U.S. EPA.)

COLLECTION AND PRETREATMENT

Groundwater samples can be collected with portable or dedicated in situ sampling equipment. Portable equipment includes bailers, syringes, suction-lift pumps, submersible pumps, and gas-driven devices. In situ sampling equipment includes cone penetrometer samplers (e.g., Hydropunch, BAT, CPT, or DMLS), chemical-sensitive probes, ion-selective electrodes, fiber-optic chemical sensors, multilevel capsule samplers, and multiport casings. A description of these types of equipment as well as their advantages and disadvantages is in the U.S. EPA desk reference guide on subsurface characterization and monitoring techniques (1993).

Selecting sampling equipment should be based on the purpose of the sampling as well as the construction materials of the sampling equipment and the method of sample delivery. The construction materials of the sampling device could affect the integrity of the sample because constituents can leach from the materials into the water samples or contaminants from the water sample can adsorb onto the sampler materials (Barcelona, Gibb, and Miller 1983). Therefore, inert materials should be specified when necessary. The method of sample delivery is important because devices that cause aeration, degassing, or pressure changes of the sample may not preserve the chemical quality of the sample. For example, devices that introduce dissolved oxygen into the sample could cause oxidation of ferrous iron to ferric iron, which affects the speciation and concentration of many chemical constituents in the sample (Hrzog, Pennimo, and Nielson 1991). Turbulence and depressurization can affect the sample's original content of dissolved oxygen, carbon dioxide, and volatile organic compounds (Barcelona, Gibb, and Miller 1983).

Another decision environmental engineers should make before sampling is whether to filter the sample in the field. This decision should be based on the characteristics of the constituents and the purpose of the sampling program. For example, samples requiring analysis for dissolved metals, alkalinity, and anionic species should be filtered. In contrast, samples for dissolved gases or volatile organics should not be filtered since the handling required by filtration could lose these chemicals. Furthermore, filtration should be performed when the sampling program is concerned only with those constituents that are dissolved in groundwater, excluding all constituents which can be adsorbed onto particulate matter in suspension, such as PCBs or polynuclear aromatic hydrocarbons. However, when a drinking water source is studied, samples should not be filtered before analysis because water taken from private wells is generally not filtered before use. In some instances, engineers must run parallel sets of filtered and unfiltered samples to determine the dissolved and adsorbed portions of the constituent of interest.

Filtration is accomplished by vacuum, pressure, or in-line filtration devices. Stolzenburg and Nichols (1986) describe a variety of filtration equipment and their effects on sampling. The preferred device is the inline filter because it reduces the aeration and degassing of the sample as well as the potential of sample cross contamination caused by improper equipment decontamination.

To prevent cross contamination, engineers should decontaminate the equipment used for sample collection or pretreatment prior to and after each use. The decontamination should involve a minimum of scraping or brushing to remove any soil or residue from the device, washing with potable or deionized water, washing with detergents or cleaning fluids such as acetone, and pressure cleaning with a high-pressure steam cleaner.

QUALITY ASSURANCE AND QUALITY CONTROL

Groundwater sampling requires a quality assurance and quality control (QA/QC) plan which is designed to minimize sources of error in each stage of the sampling process, from sample collection to analysis and reporting. The QA/QC plan should include procedures and requirements for chain-of-custody, sample storage and holding time, use of quality control samples, instrument calibration, sample analysis, laboratory validation, documentation, and record keeping.

A chain-of-custody must be filed and maintained from the moment the sample bottles are released from the laboratory until the samples are received by the laboratory. The samples must be stored in conditions that preserve their integrity. Some samples require acidification to a specified pH or cooling to a specified temperature. In addition, the recommended maximum holding time for the analyte of interest should not be exceeded. Required holding times can range from hours to days as shown in Table 4.2.1.

The purpose of quality control samples is to detect additional sources of contamination in the field or laboratory that might potentially influence the analytical values reported in the samples. Examples of quality control samples include trip blanks and field blanks.

Trip blanks consist of a set of sample bottles filled at the laboratory with laboratory demonstrated analyte-free water. Trip blanks travel to the site with the empty sample bottles, at a rate of one per shipment, and back from the site with the collected samples to simulate sample handling conditions. Contaminated trip blanks indicate inadequate bottle cleaning or blank water of questionable quality.

Field blanks serve the same purpose as trip blanks but are also used to indicate potential contamination from ambient air or sampling instruments. At the field location, analyte-free water is passed through clean sample equipment and placed in an empty sample container for analysis. Therefore, by being opened in the field and transferred over a cleaned sampling device, the field blank can indi-

TABLE 4.2.1   REQUIRED HOLDING TIMES FOR SEVERAL ANALYTES

Parameters (Type)	Volume Required (mL) 1 Sample[a]	Containers (Material)	Preservation Method	Maximum Holding Period
Well purging				
pH (grab)	50	T,S,P,G	None; field det.	<1 hr[b]
Ω^4 (grab)	100	T,S,P,G	None; field det.	<1 hr[b]
T (grab)	1000	T,S,P,G	None; field det.	None
Eh (grab)	1000	T,S,P,G	None; field det.	None
Contamination Indicators				
pH, Ω^4 (grab)	As above	As above	As above	As above
TOC	40	G,T	Dark, 4°C	24 hr
TOX	500	G,T	Dark, 4°C	5 days
Water quality				
Dissolved gases ($O_2CH_2CO_2$)	10 mL minimum	G,S	Dark, 4°C	<24 hr
Alkalinity/acidity	100	T,G,P	4°C/None	<6 hr[b] <24 hr
	Filtered under pressure with appropriate media			
(Fe, Mn, Na^+, K^+, Ca^{++}, Mg^{++})	All filtered 1000 mL[f]	T,P	Field acidified to pH <2 with HNO_2	6 months[c]
(PO_4^-, Cl_3^- Silicate)	@50	(T,P,G glass only)	4°C	24 hr[f] 7 days[c], 7 days
NO_3^-	100	T,P,G	4°C	24 hr[d]
SO_4^-	50	T,P,G	4°C	7 days[c]
OH_4^+	400	T,P,G	4°C/H_2SO_4 to pH <2	24 hr[f] 7 days
Phenols	500	T,G	4°C/H_2PO_4 to pH <4	24 hours
Drinking Water suitability As, Ba, Cd, Cr, Pb, Hg, Se, Ag	Same as above for water quality cations (Fe, Mn, etc.)[f]	Same as above	Same as above	6 months
F	Same as chloride above	Same as above	Same as above	7 days
Remaining organic	As for TOX/TOC, except where analytical parameters method calls for acidification of sample			24 hours

Source: U.S. Environmental Protection Agency, 1993, *Subsurface characterization and monitoring techniques, a desk reference guide,* Vol. 1, USEPA 625/R-93/003a, May (U.S. EPA).

T = Teflon; S = stainless steel; P = PVC, polypropylene, polyethylene; G = borosilicate glass.

[a]It is assumed that at each site, for each sampling date, replicates, a field blank, and standards must be taken at equal volume to those of the samples.

[b]Temperature correction must be made for reliable reporting. Variations greater than ±10% can result from a longer holding period.

[c]In the event that NHO₂ cannot be used because of shipping restrictions, the sample should be refrigerated to 4°C, shipped immediately, and acidified on receipt at the laboratory. Container should be rinsed with 1:1 HNO_3 and included with sample.

[d]28-day holding time if samples are preserved (acidified).

[e]Longer holding times in U.S. EPA (1986b).

[f]Filtration is *not* recommended for samples intended to indicate the mobile substance lead. See Puis and Barcelona (1989a) for more specific recommendations for filtration procedures involving samples for dissolved species.

cate ambient and equipment conditions that can potentially affect the quality of the associated samples.

—Ahmed Hamidi

References

Aller, L. et al. 1991. *Handbook of suggested practices for the design and installation of groundwater monitoring wells.* EPA/600/4-89/034.

Barcelona, M. et al. 1990. Contamination of groundwater: Prevention, assessment, restoration. *Pollution Technology Review* 184. Park Ridge, N.J.: Noyes Data Corporation.

Barcelona, M.J., J.P. Gibb, and R.A. Miller. 1983. *A guide to the selection of materials of monitoring well construction and groundwater sampling.* Illinois State Water Survey Report 327.

Campbell, M.D., and J.H. Lehr. 1975. Well cementing. *Water Well Journal* 29, no. 7:39–42.

Davis, H.E., J. Jehn, and S. Smith. 1991. Monitoring well drilling, soil sampling, rock coring, and borehole logging. In *Practical handbook of groundwater monitoring,* edited by David Nielsen. Chelsea, Mich.: Lewis Publishers.

Edge, R.W., and K. Cordy. 1989. The HydroPunch: An in situ sampling tool for collecting groundwater from unconsolidated sediments. *Groundwater Monitoring Review* (summer):177–183.

Gibb, J.P., R.M. Schuller, and R.A. Griffin. 1981. *Procedures for the collection of representative water quality data from monitoring wells.* Cooperative Groundwater Report 7, Illinois State Water and Geological Surveys.

Hrzog, B., J. Pennimo, and G. Nielson. 1991. Groundwater sampling. In *Practical handbook of groundwater monitoring,* edited by David Nielsen. Chelsea, Mich.: Lewis Publishers.

Meridith, D.V., and D.A. Brice. 1992. Limitations on the collection of representative samples from small diameter monitoring wells. Groundwater Management II (6th NOAC): 429–439.

Miller, G.D. 1982. Uptake and release of lead, chromium, and trace level volatile organics exposed to synthetic well castings. Proceedings of Second National Symposium on Aquifer Restoration and Groundwater Monitoring, 26–28 May, Columbus, Ohio: NWWA.

Nielson, D.M., and R. Schalla. 1991. Design and installation of groundwater monitoring wells. In *Practical handbook of groundwater monitoring,* edited by David Nielsen. Chelsea, Mich.: Lewis Publishers.

Seanor, A.M., and L.K. Brannaka. 1983. Efficient sampling techniques. *Groundwater Age* 17, no. 8:41–46.

Stolzenburg, T.R., and D.G. Nichols. 1986. Effects of filtration methods and sampling on inorganic chemistry of sampled well water. Proceedings of the Sixth National Symposium and Exposition on Aquifer Restoration and Groundwater Monitoring, 216–234. Dublin, Ohio: National Water Well Association.

U.S. Environmental Protection Agency. 1986. *RCRA groundwater monitoring technical enforcement guidance document.* OSWER-9950.1. Washington, D.C.: U.S. Government Printing Office.

———. 1993. *Subsurface characterization and monitoring techniques, a desk reference guide,* Vol. 1. USEPA/625/R-93/003a (May). U.S. EPA.

Unwin, J.P., and D. Huis. 1983. A laboratory investigation of the purging behavior of small diameter monitor well. Proceedings of the Third Annual Symposium on Groundwater Monitoring and Aquifer Restoration, 257–262. Dublin, Ohio: National Water Well Association.

Williams, C., and L.G. Evans. 1987. Guide to the selection of cement, bentonite and other additives for use in monitor well construction. Proceedings of First National Outdoor Action Conference, 325–343. Dublin, Ohio: National Water Well Association.

Wilson, L.G., and J.M. Dworkin. 1984. Development of a primer on well water sampling for volatile organic substances. Bk. 1, Chap. D-2 of *U.S. Geological Survey techniques of water resources investigations.*

5
Groundwater Cleanup and Remediation

5.1
SOIL TREATMENT TECHNOLOGIES

Restoration or cleanup of a contaminated aquifer usually involves also addressing the contaminated soils at the vadose zone. The residual contaminants in the vadose need to be treated or removed to prevent the continuous feed of contaminants to groundwater due to leaching, rainfall percolation, or groundwater table fluctuations. Several techniques are available to treat contaminants in the vadose zone. These techniques include excavation and removal, physical treatments, biological treatments, thermal treatments, and stabilization treatments. Selecting the appropriate method depends on the volume of soils to be handled, the type of soils and contaminants, the regulatory requirements, and costs.

Excavation and Removal

Excavation and soil removal is one of the most common activities in groundwater remediation and cleanup. Excavation involves removing contaminated soil from the unsaturated zone to prevent further groundwater contamination by the residuals present in that zone. Excavation is often used at sites where site conditions preclude onsite treatment, stabilization, or capping of the contaminated unsaturated zone. Factors affecting excavation include the volume of soils to be handled, the location of the area to be excavated, the type of soils and contaminants, and the regulatory requirements. The excavated material is often disposed of at a permitted landfill or treated and reused.

Physical Treatment

The physical treatments for treating contaminants in the vadose zone include soil–vapor extraction, soil washing, and soil flushing.

SOIL–VAPOR EXTRACTION

This treatment technology removes volatile compounds from the vadose zone. Airflow is injected through extraction wells creating a vacuum and a pressure gradient that induces volatiles to diffuse through the extraction wells. The volatiles are collected as gases and treated aboveground. The technology is effective for halogenated volatiles and fuel hydrocarbons (U.S. EPA 1991a). The technology is also cost effective when large volumes of soil are involved; since treatment takes place onsite, the risks and costs associated with transporting large volumes of contaminated soils are eliminated.

The technology, however, is less effective in soils with low air permeability, low temperatures, or high carbon content. In addition, although this technology reduces the volume of the contaminants, the toxicity of the contaminants is not reduced.

SOIL WASHING

Soil washing removes adsorbed contaminants from soil particles. The process involves excavating the contaminated soil and washing it with a leaching agent, a surfactant, or chelating agency or adjusting the pH (U.S. EPA 1990c). Sometimes extraction agents are added to enhance the process. The process reduces the volume of contaminant; however, residual suspended solids and sludges from the process may need further treatment since they contain a higher concentration of contaminant than the original. The technology is effective for halogenated semivolatiles, fuel hydrocarbons, and inorganics (U.S. EPA 1993a).

The technology, however, is less effective when the soil contains a high percentage of silt and clay particles or high organic content. In addition, this technology reduces the volume of the contaminants, but the toxicity of the contaminants is unchanged.

SOIL FLUSHING

Soil flushing is an in situ process whereby environmental engineers apply a water-based solution to the soil to enhance the solubility of the contaminant (U.S. EPA 1991b). The water-based solution is applied through injection wells or shallow infiltration galleries. The contaminants are mobilized by solubilization or through chemical reactions with the added fluid. The generated leachate must be intercepted by extraction wells or subsurface drains and pumped to the surface for aboveground treatment. The technology is effective for nonhalogenated volatile organics and for soils with high permeability.

The technology, however, is less effective for soils with low permeability or with particles that strongly adsorb contaminants such as clays. In addition, special precautions are necessary to prevent groundwater contamination.

Biological Treatment

Slurry biodegradation, ex situ bioremediation and land farming, and in situ biological treatment are biological treatments for treating contaminants in the vadose zone. These treatment are discussed next.

SLURRY BIODEGRADATION

The slurry biodegradation process involves excavating the contaminated soil and mixing it in an aerobic reactor with water and nutrients. This process maximizes the contact between the contaminants and the microorganisms capable of degrading those contaminants. The temperature in the reactor is usually maintained at an appropriate level, and neutralizing agents are often added to adjust the pH to an acceptable range (U.S. EPA 1990b). After the treatment is complete, the slurry is dewatered, and the soil can be redeposited on site. This technology is effective for soils contaminated with fuel hydrocarbons (U.S. EPA 1993a). In addition, the contaminants can be completely destroyed and the soil reused.

The technology, however, is less effective for contaminants with low biodegradability. In addition, the presence of chlorides or heavy metals as well as some pesticides and herbicides in the soil can reduce the effectiveness of the process by inhibiting the microbial action.

EX SITU BIOREMEDIATION AND LANDFARMING

This process involves excavating the contaminated soil, piling it in biotreatment cells, and periodically turning it over to aerate the water (U.S. EPA 1993a). The moisture, heat, nutrients, oxygen, and pH are usually controlled in the process. In addition, volatile emissions as well as leachate from the biotreatment cells should be controlled. The technology is effective for soils contaminated with fuel hydrocarbons. Also, the contaminants can be completely destroyed and the soil reused.

IN SITU BIOLOGICAL TREATMENT

This process enhances the naturally occurring biological activities in the contaminated subsurface soil. Circulating either a nutrient and oxygen-enriched water-based solution or a forced air movement which provides oxygen in the soil enhances the naturally occuring microbes (U.S. EPA 1991a). In the latter process, also called *bioventing*, the air flow rate is lower than in vapor extraction since the objective is to deliver oxygen while minimizing volatilization and the release of contaminants to the at-

mosphere. The technology is effective for nonhalogenated volatiles and fuel hydrocarbons. In addition, the contaminant toxicity is reduced or even eliminated. The technology, however, is less effective for nonbiodegradable compounds and for soils with low permeability.

Thermal Treatment

Thermal treatment is used to treat contaminants in the vadose zone and includes incineration and thermal desorption. A brief description of these processes follows.

INCINERATION

Incineration is a process whereby organic compounds in contaminated soil are destroyed in the presence of oxygen at high temperatures (U.S. EPA 1990a). Three common types of incinerators are rotary kilns, circulating fluidized beds, and infrared incinerators. The excavated contaminated soil is fed into the incinerator and incinerated at temperatures ranging from 1600 to 2200°F. Because the residual ash may contain residual metals, it must be disposed of in accordance with appropriate regulations. In addition, the generated flue gases must be handled with appropriate air pollution control equipment.

Incineration is potentially effective for halogenated and nonhalogenated volatiles as well as fuel hydrocarbons and pesticides. Most organic contaminants are destroyed by this technology; however, metals are not destroyed and end up in the flue gases or the ash. In addition, certain types of soils such as clay soils or soil containing rocks may need screening prior to incineration.

THERMAL DESORPTION

Thermal desorption is a physical separation process in which the excavated contaminated soil is heated to a temperature at which the water and organic contaminants are volatilized (U.S. EPA 1991d). The volatilized contaminants are then sent to a gas treatment system. Low-temperature thermal desorption is potentially effective for halogenated semivolatiles, nonhalogenated volatiles, and pesticides (U.S. EPA 1993a). High-temperature thermal desorption is effective for halogenated volatiles and semivolatiles as well as fuel hydrocarbons.

The contaminants, however, are not destroyed by this technology and require further gas treatment. In addition, the technology is less effective for tightly aggregated soils or those containing large rock fragments.

Stabilization and Solidification Treatment

Stabilization and vitrification treatments are also used to treat contaminants in the vadose zone. These treatments are described next.

STABILIZATION

The soil stabilization process can be used in either in situ or ex situ treatment. The process involves mixing the contaminated soil with binding materials such as cement, lime, or thermoplastic binders to bind the contaminants to the soil and reduce their mobility (U.S. EPA 1993b). Depending on the process and binding material, the final product ranges from a loose, soil-like material to concrete-like molded solids. Pretreatment is usually required for soils with high contents of oil and grease, surfactants, or chelating agents. The process is effective for soils, sludges, or slurries contaminated with inorganics.

The technology, however, is not effective for soils contaminated with organics or soils with high water or clay content. Organics, sulfates, or chlorides can interfere with the curing of the solidified product. Clay can interfere with the mixing process, adsorbing the key reactants and interrupting the polymerization chemistry of the solidification agents. Furthermore, the stabilization process increases the volume of treated soil since reagents are added.

VITRIFICATION

Soil vitrification is used in both in situ and ex situ treatment. The process involves inserting large graphite electrodes into the soil and applying a high current of electricity to the electrodes (U.S. EPA 1992). The electrodes are typically arranged in 30-foot squares and connected by graphite on the soil surface. The heat causes a melt that gradually works downward through the soil incorporating inorganic contaminants into the melt and paralyzing organic components. After the process is complete and the ground has cooled, the fused waste material is dispersed in a chemically inert, stable, glass-like product with low leaching characteristics.

The technology is potentially effective for halogenated and nonhalogenated volatiles and semivolatiles as well as fuel hydrocarbons, pesticides, and inorganics. The process reduces the mobility of the contaminants, and the vitrified mass resists leaching for geological time periods. The technology, however, is energy-intensive, and the off-gases must be collected and treated before release.

—Ahmed Hamidi

References

U.S. Environmental Protection Agency. 1990a. *Mobile/transportable incineration treatment*. EPA/540/2-90/014. Washington, D.C.: U.S. EPA.

———. 1990b. *Slurry biodegradation*. EPA/540/2-90/016 (September). Washington, D.C.: U.S. EPA.

———. 1990c. Soil washing treatment. *Engineering Bulletin* EPA/540/2-90/017 (September). Washington, D.C.: U.S. EPA.

———. 1991a. *Bioremediation in the field*. EPA/540/2-91/018. Washington, D.C.: U.S. EPA.

———. 1991b. *In situ soil flushing*. EPA/540/2-91/021. Washington, D.C.: U.S. EPA.

———. 1991c. *In-situ soil vapor extraction treatment*. EPA/540/2-91/006 (May). Washington, D.C.: U.S. EPA.

———. 1991d. *Thermal desorption treatment*. EPA/540/2-91/008 (May). Washington, D.C.: U.S. EPA.

———. 1992. *Vitrification technologies for treatment of hazardous and radioactive waste*. EPA/625/R-92/002 (May). Washington, D.C.: U.S. EPA.

———. 1993a. *Remediation technologies screening matrix and reference guide*. EPA/542/B-93/005 (July). Washington, D.C.: U.S. EPA.

———. 1993b. *Solidification/stabilization of organics and inorganics*. EPA/540/S-92/015 (May). Washington, D.C.: U.S. EPA.

5.2
PUMP-AND-TREAT TECHNOLOGIES

Pump-and-treat systems consist of a groundwater withdrawal system and an aboveground treatment system. The groundwater withdrawal system, also called the *containment system*, includes pumping wells or subsurface drains designed to remove the contaminants from the groundwater system and control the plume from further migration. In some cases, injection wells are used to inject treated water back into the aquifer. Aboveground treatment systems include chemical, physical, and biological treatment technologies.

Withdrawal and Containment Systems

As previously stated, the withdrawal and containment systems include well systems and subsurface drains. A description of these systems follows.

WELL SYSTEMS

Well systems remove contaminants from groundwater and stop the plume from further migration by manipulating the subsurface hydraulic gradients. Three general classes of well systems are well points, deep wells, and injection wells. Well points use suction lifting as the standard technique for pumping water; therefore, they can be used only for shallow aquifers where the suction lifting is less than 25 feet. Figure 5.2.1 shows several closely spaced point wells connected to a centrally located suction lift pump through a single main header pipe. Deep-well systems are used for greater depths and are usually pumped individually by submersible pumps. Dual pumps are used for floating product recovery as shown in Figure 5.2.2. In injection wells, the injection of clean or treated water into the aquifer flashes the aquifer or forms a barrier to groundwater flow.

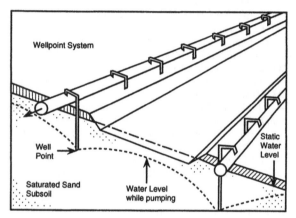

FIG. 5.2.1 Suction lift and a series of point wells. (Reprinted from S. Sommer and J.F. Kichens, 1980, *Engineering and development support of general decon technology for the DARCOM installation and restoration program, Task 1: Literature review on groundwater containment and diversion barriers*, Draft report by Atlantic Research Corp. to U.S. Army Hazardous Materials Agency, Contract No. DAK 11-80-C-0026, [October], Aberdeen Proving Ground.)

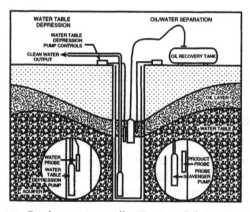

FIG. 5.2.2 Dual pumping wells. (Reprinted from E.K. Nyer, 1992, *Groundwater treatment technology*, 2d ed., New York: Van Nostrand Reinhold.)

Design Considerations

The design of a well system involves determining the number of wells needed, placing and spacing the wells, and determining the pumping cycles and rates of the wells. The number and spacing of the wells should completely capture the plume of contamination and produce as little uncontaminated water as possible to reduce treatment costs. In addition, the well's capture zones should intersect each other to prevent dead spots where contaminants stay stagnant or routes where the contaminant can escape the zone of capture. Environmental engineers determine the zone of capture by plotting the drawdowns within the radius of influence of each well on the potentiometric surface map of the site and calculating the cumulative drawdowns. The radius of influence of each well is determined by pumping test analysis as discussed in Section 2.3 or estimated from the following formulas when pumping test data are lacking (Kuffs et al. 1983):

Equilibrium:

$$R_0 = 3(H - hw)(0.47K)^{1/2} \qquad 5.2(1)$$

Nonequilibrium:

$$R_0 = r_w + (Tt/4790 \ S)^{1/2} \qquad 5.2(2)$$

where:

R_0 = radius of influence, ft
K = permeability, gpd/ft^2
H = total head, ft
h_w = head in well, ft
Q = pumping rate, gpm
r_w = well radius, ft
T = transmissivity, gpd/ft
t = time, min
S = storage coefficient, dimensionless

Methods of Construction

The construction of a well system involves setting up the drilling equipment, drilling the well hole, installing casings and liners, grouting and sealing annular spaces, installing well screens and fittings, packing gravel and placing material, and developing the well. Detailed discussions on these aspects can be found in Johnson Division, UOP, Inc. (1975). In addition, Figure 5.2.3 shows typical well construction detail. In recent years, several innovative well installation techniques have been developed including installing horizontal wells which act as subsurface drains but require less soil excavation and disturbances (Oakley et al. 1994).

Operation and Maintenance

Equilibrium pumping is often used for plume management; however, nonequilibrium pumping has advantages in cases of floating and sinking plumes and can be used to flush sorbed contaminants associated with the residual phase.

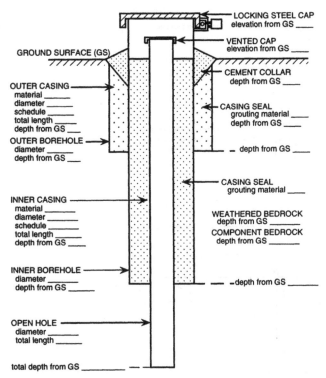

FIG. 5.2.3 Typical well installation.

Pulsed pumping of recovery wells can be used to washout residuals from unsaturated zones, allow contaminants to diffuse out of low permeability zones, and flush and bring stagnant zones into active flow paths. Pulsed pumping, however, incurs additional costs and concerns that must be evaluated for site specific conditions (U.S. EPA 1989).

Cost

The costs of well systems vary from site to site depending on the geology, the depth of the aquifer, the extent and type of contamination, the periods and durations of pumping, and the electrical power requirements. A cost analysis study for a variety of well systems can be found in Cambel and Lehr (1977) and in Powers (1981).

Advantages and Limitations

Well system technology is an efficient and effective means of assuring groundwater pollution control. Wells can be readily installed, or previously installed monitoring wells can sometimes be used as part of a well system. In addition, the technology provides high-design flexibility, and the construction costs can be lower than artificial barriers. However, wells require continued maintenance and monitoring after installation; therefore, operation and maintenance costs can be high. In addition, the application of this technology to fine soils is limited due to the low yield and small radius of influence in these soils.

SUBSURFACE DRAINS

Subsurface drains involve excavating a trench and placing a perforated pipe and coarse material such as gravel in the trench. The drain usually drains by gravity to a sump where the water is pumped to the surface for treatment. Subsurface drains essentially function like an infinite line of extraction wells, creating a continuous zone of depression in which groundwater flows towards the drain. Two types of subsurface systems are relief drains and interceptor drains. The major difference between these drains is that the drawdown created by an interceptor drain is proportional to the hydraulic gradient, whereas the drawdown created by a relief drain is a function of the hydraulic conductivity and depth to the impermeable layer below the drain.

Environmental engineers use *relief drains* primarily to lower the water table and prevent its contact with waste material or to contain a plume in place and prevent contamination from reaching a deeper aquifer. Relief drains can be installed in parallel on either side of a waste site or completely around the perimeter of the waste site as shown in Figure 5.2.4. The areas of influence of relief drains should overlap to prevent the contaminated groundwater from escaping between the drain lines.

Engineers use *interceptor drains* to intercept a plume hydraulically downgradient from its source and prevent the contamination from reaching wells and surface water located downgradient from the site. Interceptor drains are installed perpendicular to groundwater flow and downgradient of the plume of contamination as shown in Figure 5.2.5. In some cases, engineers use interceptor drains in conjunction with a barrier wall to prevent infiltration of clean water from downgradient of the drain thereby reducing treatment costs (see Figure 5.2.6). A series of interceptor drains or collector pipes (laterals) can be connected to a main pipe (header) as shown in Figure 5.2.7.

Design Considerations

The primary design components of a subsurface drain system are (1) the location of the drains, (2) the spacing of the drains, (3) the pipe diameter and slope, and (4) the envelope and filter materials around the pipe.

An interceptor drain should be installed perpendicular to the groundwater flow direction and downgradient from the plume of contamination. The drain should be installed on top of a layer of low hydraulic permeability to prevent underflow beneath the drain. The location of the drain should be selected so that the upgradient and downgradient influences of the drain completely capture the contamination plume. The upgradient and downgradient influences of an interceptor drain can be calculated using the following equations described by Van Hoorn and Vandemolen (1974) and Kuffs (1983):

$$D_u = 1.33m_sI \qquad \text{5.2(3)}$$

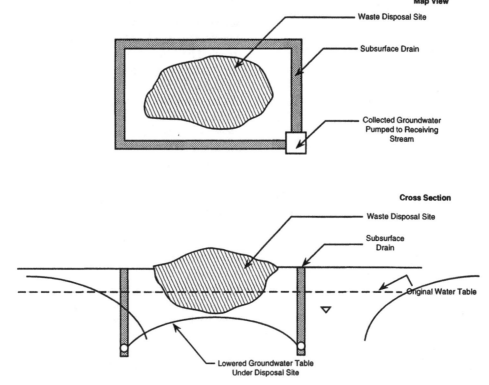

FIG. 5.2.4 Relief drains around the perimeter of a waste site. (Reprinted from U.S. Environmental Protection Agency, 1985, *Leachate plume management*, EPA/540/2-85/004, Washington, D.C.: U.S. EPA.)

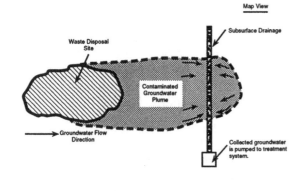

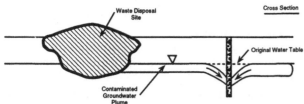

FIG. 5.2.5 Interceptor drains downgradient of the plume of contamination. (Reprinted from U.S. Environmental Protection Agency, 1985, *Leachate plume management*, EPA/540/2-85/004, Washington, D.C.: U.S. EPA.)

$$D_d = \frac{KI}{Q} \cdot (d_e - h_d - D_2) \qquad 5.2(4)$$

where:

D_u = effective distance of drawdown upgradient, ft
m_s = saturated thickness of aquifer not affected by drainage, ft
I = hydraulic gradient
D_d = downgradient influence, ft
K = hydraulic conductivity, ft/day
Q = drainage coefficient, ft/day
d_e = depth of drain, ft
h_d = depth of drawdown, ft
D_2 = distance from ground surface to water table prior to drainage at the distance D_d downgradient from the drain, ft

The spacing between two parallel relief drains should be selected so that their combined drawdown is adequate to lower the water table beneath the waste. The minimum spacing, however, is often imposed by the boundaries of the waste material. The drain spacing depends on the hydraulic conductivity of the aquifer, the depth of the impermeable layer beneath the drain, the cross-sectional area of the drain, the water level in the drain, and precipitation and other sources of recharge. The spacing between two parallel drains resting on an impermeable barrier can be

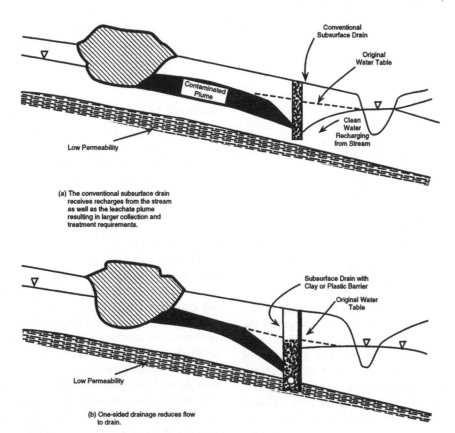

(a) The conventional subsurface drain receives recharges from the stream as well as the leachate plume resulting in larger collection and treatment requirements.

(b) One-sided drainage reduces flow to drain.

FIG. 5.2.6 Interceptor drains in conjunction with a barrier wall. (Reprinted from U.S. Environmental Protection Agency, 1985, *Leachate plume management,* EPA/540/2-85/004, Washington, D.C.: U.S. EPA.)

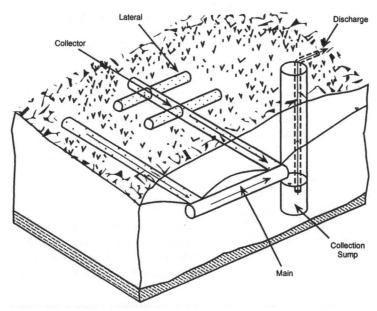

FIG. 5.2.7 Interceptor drains connected to a header. (Reprinted from U.S. Environmental Protection Agency, 1985, *Leachate plume management,* EPA/540/2-85/004, Washington, D.C.: U.S. EPA.)

calculated with the use of the Wasseling (1973) equation as

$$L = \left(\frac{8KDH + 4KH^2}{Q}\right)^{0.5} \qquad 5.2(5)$$

where:

L = drain spacing, ft
K = hydraulic conductivity, ft/sec
D = distance between the water level in the drain and the impermeable layer, ft
H = height of the water table above the water level in the drain midway between the two drains, ft
Q = drain drainage rate per unit surface area, ft/sec

For a two-layered soil, Hooghoudt, as described by Wasseling (1973), developed a modified equation of

$$L = \left(\frac{8K_1DH + 4K_2H^2}{Q}\right)^{0.5} \qquad 5.2(6)$$

where K_1 and K_2 are the hydraulic conductivities above and below the drain, and d is the equivalent depth of the aquifer below the drain as illustrated in Figure 5.2.8. Using this equation involves either a trial-and-error procedure or the use of monographs which have been developed specifically for equivalent depth and drain spacing (U.S. EPA 1985b; Repa et al. 1982). Other equations for different subsurface configurations are available in Cohen and Miller (1983).

The diameter of the pipe can be calculated with the use of Manning's equation, assuming that the carrying capacity of the pipe is equal to the design seepage. The resulting equation (Luthin 1957) is

$$d = 0.892(q \cdot A)^{0.375} \cdot A^{-0.1875} \qquad 5.2(7)$$

where:

d = inside diameter of pipe, in
A = drainage area, acres
q = seepage coefficient, in/day
I = hydraulic gradient

The slope of the pipe should be selected so that the flow velocity in the drain is greater than the critical velocity of siltation of the soil that enters the drain (Soil Conservation

Service 1973). When the velocity is less than 1.4 ft/sec, filter fabrics and silt traps or cleanouts should be installed around the pipe and along the subsurface drain.

Methods of Construction

The construction of a subsurface drain involves trench excavation, dewatering, wall stabilization, pipe installation, and backfilling. Trench excavation is the most significant step in the construction of a subsurface drain. A variety of excavation equipment can excavate the trench; the optimum is determined by the depth, width, length of the trench, and the type of material being excavated. Dewatering can be performed by open pumping, predrainage using well points, or groundwater cutoff. Wall stabilization methods include shoring for deep excavations or open cuts for shallow excavations. Continuous trenching machines can accomplish all excavation and pipe installation operations simultaneously (Oakley et al. 1994); however, this machinery is limited to small diameter and relatively shallow subsurface drains.

Another important aspect in subsurface drain installation is the placement of filter and envelope materials around the pipe to prevent soil particles from entering and clogging the pipe. Geotextiles and well-graded sand and gravel can be used as filter materials. The general requirement for envelopes is that their hydraulic conductivity is higher than that of the base material. Design procedures for filters and envelopes are in the Soil Conservation Service (1973).

Operation and Maintenance

Subsurface drains require frequent inspection and maintenance during the first year or two of operation. Typical problems that can develop in drainage systems and require maintenance include clogging of the drain or manhole by sediment buildup or buildup of chemical compounds such as iron and manganese. Clogged pipes can be cleaned by hydraulic jetting, mechanical scrapping, or chemical treatment in cases of chemical buildup.

Cost

Costs of subsurface drain systems vary from site to site depending on the geology, the depth of the aquifer, the extent and type of contamination, the periods and durations of pumping, and the electrical power requirements. The major costs of a subsurface drain system occur during system installation. These costs include excavation, dewatering, pipe bedding, filter and envelop materials, pipes, manholes, and pumps. Typical costs for subsurface drain installation are in Means (1994).

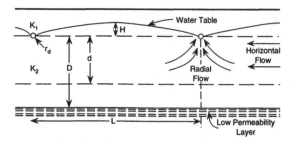

FIG. 5.2.8 Subsurface drain formulation. (Reprinted from U.S. Environmental Protection Agency, 1985, *Leachate plume management,* EPA/540/2-85/004, Washington, D.C.: U.S. EPA.)

Advantages and Limitations

For shallow contaminations, subsurface drains are more cost-effective than wells particularly in aquifers with low or variable permeability. Construction methods for subsurface drains are simple, and operation costs are relatively low since flow to the underdrain is by gravity. In addition, subsurface drains provide considerable design flexibility since adjusting the depth or modifying the envelope material can alter the spacing to some extent. However, subsurface drains are not suited to poorly permeable soils, to deep contaminant plumes, or beneath existing sites. In addition, subsurface drains require continuous and careful monitoring to assure adequate leachate collection and prevent pipe clogging.

Treatment Systems

The most commonly used treatment in pump-and-treat technologies is physical treatment. Physical treatment includes density separation, filtration, adsorption, air stripping, and reverse osmosis. Each of these processes can be used individually or in conjunction with others (e.g., treatment trains) as shown in Figure 5.2.9. In addition, most treatment systems include equalization and spill control to protect the treatment works from shock pollutants and hydraulic loadings. The selection of the process is usually based on the type of contaminant, influent concentration, effluent requirements, and cost.

DENSITY SEPARATION

Density separation is a process whereby the water and contaminant are separated based on their individual densities. This treatment is often used for the pretreatment of suspended solids or floating immiscible products that could be present in pumped groundwater. For suspended solids, the most commonly used equipment is clarifiers, settling chambers, and sedimentation basins. For immiscible products, such as oil and grease, the most commonly used equipment is oil–water separators. Both suspended solids and oil and grease must generally be removed from contaminated groundwater prior to further treatment because these materials can foul instruments and interfere with other processes. Furthermore, oil and grease and suspended solids can damage the environment and cause a significant pollution problem to the receiving body of water.

The most common oil–water separators are the American Petroleum Institute (API) gravity separators and the parallel-plate separators. The design of an oil–water separator is based on the amount of oil present in the water, the oil droplet size distribution, the presence of surfactants, the specific gravity of the oil, and the water temperature. A step-by-step procedure for the design of an oil–water separator is in Corbitt (1990). Once the oil or floating product is at the surface, it can be removed from the water by slotted pipes, dip tubes, or belt or rope skimmers.

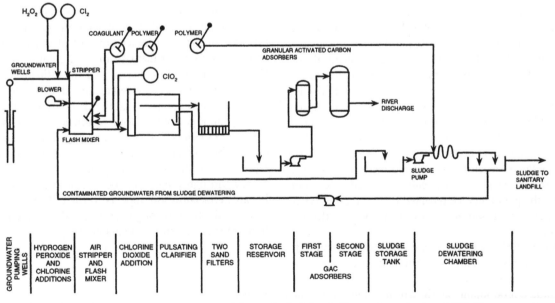

FIG. 5.2.9 Typical treatment trains. (Reprinted from North Atlantic Treaty Organization, 1993, *Demonstration of remedial action technologies for contaminated land and groundwater,* Vol. 2, pt. 2, no. 190, EPA/600/R-93/012, NATO Committee on the Challenges of Modern Society.)

FILTRATION

Filtration is a process whereby suspended solids are removed from the influent by forcing the water through a filter of porous medium such as sand or sand with anthracite or coal. The purpose of filtration is to reduce the concentration of suspended solids, such as carbon columns, prior to certain treatment processes. The most common filter is a dual-media system with a layer of anthracite over a layer of sand. This filter provides better suspended solids removal with longer filter runs at higher flow rates than the more conventional single-medium filter (Corbitt 1990). The design of filters is based on the flow rate, flow scheme, and the type of medium used in the filter as discussed in Chapter 8. Up to 75% of suspended solids can be removed by dual-media filters operating at flow rates ranging from 2 to 8 gpm/ft (Oakley et al. 1994), bed depths of 24 to 48 inches, sand to anthracite ratios of 1:1 to 4:1, and a filter run of 8 to 148 hours (Corbitt 1990).

Filtration is a reliable and effective means of removing low levels of solids provided that the solid content does not vary greatly. Also, periodic filter backwashing is necessary to remove collected materials from the media. Typical backwash flow rates are 15 to 25 gpm/ft (Oakley et al. 1994) for eight to ten minutes (Corbitt 1990). The spent backwash water can be routed to the plant's headworks or to an intermediate process which provides settling.

CARBON ADSORPTION

In adsorption, the molecules of a dissolved contaminant become attached to the surface of a solid adsorbent. The most widely used adsorbent is granular activated carbon (GAC) because its porous structure provides a relatively large surface area per unit volume (1000–2000 m²/g). Collection of the molecules on the surface of the adsorbent is due to chemical or physical forces. Chemical adsorption is due to actual chemical bonding at the solid's surface. Physical adsorption is due to van der Waals' forces, which are weak bonds compared to chemical adsorption. However, because of the weak nature of these bonds, adsorbed molecules can be easily removed with a change in the solute concentration or the addition of enough energy (regeneration) to overcome the bonds. This capacity to remove certain molecules adsorbed on carbon and, thus, the possibility for repeated carbon reuse is what allows activated carbon adsorption to be a cost-effective technology.

Environmental engineers commonly use carbon adsorption to remove organic contaminants from water or air; however, they also use it to remove a limited number of inorganic contaminants as shown in Table 5.2.1. The effectiveness of GAC depends on the molecular weight, structure, and solubility of the contaminant as well as the properties of the carbon, the water temperature, and the presence of impurities such as iron and manganese. The influence of each of these parameters on the absorbability of organic contaminants is shown in Table 5.2.2. As shown in this table, carbon adsorption is suitable for high molecular weight and low solubility and polarity compounds (U.S. EPA 1988), such as chlorinated hydrocarbons, organic phosphorous, carbonates, PCBs, phenols, and benzenes. GAC can also be used in conjunction with other treatment technologies. For example, GAC can be used to treat the effluent water or offgas from an air stripper (Crittenden 1988).

Design Considerations

The most important variables in designing carbon treatment systems are the contact time and the carbon usage rate, both of which depend on the flow rate, type of contaminant, and influent and effluent concentrations. The contact time is the time allowed for the pollutant to react with the carbon exterior and enter and react with the surface of the interior pores. The contact time is the result of dividing the volume of carbon by the flow rate. The carbon usage rate is the result of dividing the volume of carbon online by the volume of water treated when the required effluent concentration is exceeded (i.e., breakpoint). The goal of the design is to find the optimum contact time which provides the lowest carbon usage rate. Typical design parameters for carbon adsorption are shown in Table 5.2.3.

The contact time and carbon usage rate for a compound are usually determined through laboratory testing. A common test method is the bed depth service time (BDST) analysis, also called the dynamic column test study (Adams and Eckenfelder 1974). In this test method, three to four columns are connected in series as shown in Figure 5.2.10. Each column is filled with an amount of carbon which provides superficial contact times of fifteen to sixty minutes per column. Effluent from each column is analyzed for the chemicals of concern, and the effluent-to-influent concentration ratio is plotted against the volume of water treated by each column. Figure 5.2.11 shows an example of a dynamic test where four columns are used and each column represents fifteen minutes of contact time T_c. The curves obtained are called *breakthrough curves* since they represent the amount of contaminated water that has passed through the carbon bed before the maximum allowable concentration appears in the effluent.

Once the breakthrough curves are determined, the carbon usage rates can be calculated as:

$$q_c = \frac{V_c}{V_w} \qquad \text{5.2(8)}$$

where:

q_c = carbon usage rate, lb/gal
V_c = volume of carbon, lb
V_w = volume of water treated when the required effluent concentration is exceeded, gal

TABLE 5.2.1 POTENTIAL FOR REMOVAL OF INORGANIC MATERIAL BY ACTIVATED CARBON

Constituents	Potential for Removal by Carbon
Metals of high sorption potential	
Antimony	Highly sorbable in some solutions
Arsenic	Good in higher oxidation states
Bismuth	Very good
Chromium	Good, easily reduced
Tin	Proven very high
Metals of good sorption potential	
Silver	Reduced on carbon surface
Mercury	CH_3HgCl sorbs easily Metal filtered out
Cobalt	Trace quantities readily sorbed, possibly as complex ions
Zirconium	Good at Low pH
Elements of fair-to-good sorption potential	
Lead	Good
Nickel	Fair
Titanium	Good
Vanadium	Variable
Iron	FE^{3+} good, FE^{2+} poor, but may oxidize
Elements of low or unknown sorption potential	
Cooper	Slight, possibly good if complexed
Cadmium	Slight
Zinc	Slight
Beryllium	Unknown
Barium	Very low
Selenium	Slight
Molybdenum	Slight at pH 6–8, good as complex ion
Manganese	Not likely, except as MnO_4
Tungsten	Slight
Free halogens	
F_2 fluorine	Will not exist in water
Cl_2 chlorine	Sorbed well and reduced
Br_2 bromine	Sorbed strongly and reduced
I_2 iodine	Sorbed very strongly, stable
Halides	
F, flouride	May sorb under special conditions
Cl^-, Br^-, I^-	Not appreciably sorbed

Source: U.S. Environmental Protection Agency, 1985, *Handbook, remedial action at waste disposal sites*, EPA/625/6-85/006 (Washington, D.C.: U.S. EPA).

Carbon usage rates are then plotted for each contact time as shown in Figure 5.2.12. The optimum contact time t_{copt} is determined as the time which provides the lowest carbon usage rate. The optimum volume of carbon bed needed is calculated as

$$V_{copt} = T_{copt} \cdot Q \qquad 5.2(9)$$

where:

V_{copt} = optimum volume of carbon bed, lb
Q = flow rate, gal

The optimal tradeoff point between a lower carbon usage rate and a smaller carbon bed size can be found through analysis. A typical minimum contact time for gasoline contaminants is fifteen minutes. This contact time corresponds to a liquid loading rate of 2 gpm/ft^2 in a standard 20,000-lb and 10-ft-diameter carbon vessel (Noonan

TABLE 5.2.2   SUMMARY OF INFLUENCE OF CONTAMINANT PROPERTIES ON THE ABSORBABILITY OF ORGANICS

Parameter	Influence on Absorbability
Molecular weight	High molecular-weight compounds adsorb better than low molecular-weight compounds.
Solubility	Low-solubility compounds are adsorbed better than high-solubility compounds.
Structure	Nonpolar compounds adsorb better than polar compounds.
	Branched chains are usually more adsorbable than straight chains.
	Large molecules are more adsorbable than small molecules.
Substituent group	*Hydroxyl* generally reduces absorbability.
	Amino generally reduces absorbability.
	Carbonyl effect varies according to host molecule.
	Double bonds effect varies.
	Halogens effect varies.
	Sulfonic usually decreases absorbability.
	Nitro often increases absorbability.
Temperature	Adsorptive capacity decreases when the water temperature increases.
Properties of carbon	Adsorption is directly proportional to the surface area of the carbon used.
	Virgin carbon has more adsorptive capacity than regenerated carbon.
Other	Iron and manganese (if present at significant levels in the water) can precipitate onto the carbon, clog its pores, and cause rapid head loss.
	Biological growth on the surface of the carbon can enhance the removal efficiency and increase the carbon service life. If the growth is excessive, however, it can clog the carbon bed.
	Excessive amounts of suspended solids (above 50 ppm) or oil and grease (above 10 ppm) can affect the efficiency of the carbon.

Source: U.S. Environmental Protection Agency, 1985, *Handbook, remedial action at waste disposal sites,* EPA/625/6-85/006 (Washington, D.C.: U.S. EPA).

and Curtis 1990). Table 5.2.4 lists the contact times as well as carbon usage rates for several organics.

Methods of Construction

GAC is available from a number of suppliers in vessels of different sizes. The vessels are typically open-top, cylindrical steel tanks for gravity systems and closed-top, cylindrical steel tanks for pressure systems. Gravity systems are operated like sand filters and are generally used for high flows, such as at municipal wastewater treatment plants. Pressure systems are generally used for smaller flows and allow higher surface loading rates (5–7 gpm/ft² compared to 2–4 gpm/ft² for gravity systems) and pressure discharge to the distribution system, saving pumping costs (Noonan and Curtis 1990).

Activated carbon is commonly made from coal; other materials such as coconut shells, lignite, wood, tires, and pulp residues can also be used. In the formation of GAC, the material is subjected first to a high temperature to remove water and other vapors from it. Then, a superheated steam is released into the material (activation) to enlarge the pores and remove ashes from it (Noonan and Curtis 1990).

TABLE 5.2.3 TYPICAL DESIGN PARAMETERS FOR CARBON ADSORPTION

Parameters	Requirements
Contact time	Generally 10–50 min; may be as high as 2 hours for some industrial wastes
Hydraulic load	2–15 gpm/ft² depending on type of contact system; see Table 5.2.1
Backwash rate	Rates of 20–30 gpm/ft² usually produce 25–50% bed expansion
Carbon loss during regeneration	4–9% 2–10%
Weight of COD removed per weight of carbon	0.2–0.8
Carbon requirements PCT plant Tertiary plant	500–1800 lb/10⁶ gal 200–500 lb/10⁶ gal
Bed depth	10–30 ft

Source: U.S. Environmental Protection Agency, 1985, *Handbook, remedial action at waste disposal sites,* EPA/625/6-85/006 (Washington, D.C.: U.S. EPA).

Operation and Maintenance

Activated carbon systems can be operated as upflow, expanded-bed columns or downflow, fixed-bed columns. Upflow expanded beds can tolerate higher suspended-solids loading than downflow beds. They also make efficient use of the carbon since fully exhausted carbon can be removed from the bottom of the bed while fresh carbon is added to the top. In addition, carbon beds can be operated in parallel (single-stage) or in series (multiple-stage) as shown in Figure 5.2.13. When operated in series, the leading contactor removes the majority of the contamination, while the second contactor removes any residual organics from the water. Furthermore, multiple-stage use allows a contactor to be completely exhausted before regeneration, while effluent quality remains protected by the subsequent contactor. When operated in parallel, contactors should stagger startup to permit bed-by-bed regeneration without reducing effluent quality.

When the adsorption capacity of the carbon is exhausted, the spent carbon can either be disposed of at a disposal site, regenerated, or reactivated for reuse. Offsite disposal at a landfill or an incinerator is the preferred method when the amount of carbon is small. For disposal at a landfill, testing and classifying the spent carbon are necessary to ensure that all regulations for disposal are being met. Spent carbon may be considered hazardous waste and may need to be disposed of at a hazardous waste landfill or burned at an incinerator where both the carbon and the hazardous waste are destroyed.

If the amount of spent carbon is large or the user has access to an offsite, multiuser facility, regeneration or reactivation for reuse may be the preferred solution. Regeneration exposes the spent carbon to steam to desorb the contaminants. Reactivation is conducted in electrical or multiple-heart furnaces where the temperature is high enough (up to 1800°F) to thermally destroy the contaminants and reactivate the carbon. Regeneration and reactivation can incur a 10 to 20% material loss and can change the adsorptive properties of the virgin grade material.

Cost

The capital costs of a GAC system include the costs of carbon, carbon vessels, pumps and piping, electrical equipment and controls, housing, design, and contingencies. The cost depends on the flow rates, type of contaminant, concentrations, and discharge requirements. Costs can vary

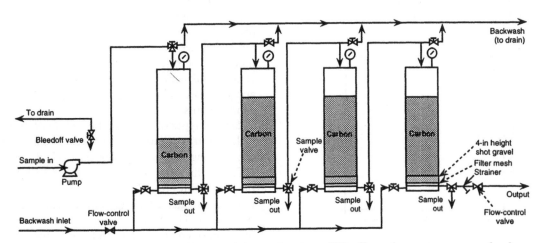

FIG. 5.2.10 Dynamic column test. (Reprinted from E.K. Nyer, 1992, *Groundwater treatment technology,* 2d ed., New York: Van Nostrand Reinhold.)

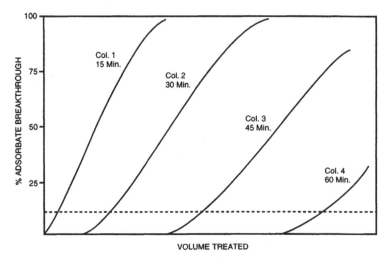

FIG. 5.2.11 Dynamic column test results breakthrough curves. (Reprinted from E.K. Nyer, 1992, *Groundwater treatment technology,* 2d ed., New York: Van Nostrand Reinhold.)

from $0.10–1.50/1000 gal treated for flow rates of 100 mgd to $1.20–6.30/1000 gal treated for flow rates of 0.1 mgd (O'Brien 1983).

Operation and maintenance costs include labor, energy, carbon replacement, and sampling and monitoring. The major cost, however, is carbon replacement which is a function of the carbon usage rate. Typical carbon costs range from $0.60 per pound for regenerated carbon to $0.75 per pound for virgin, high-quality carbon (Noonan and Curtis 1990).

Advantages and Limitations

Carbon adsorption is an effective and simple treatment technology for volatile organic compounds. In addition, GAC can be used in conjunction with other treatment technologies.

However, GAC is not recommended for low-molecular-weight and high-polarity compounds. In addition, high-

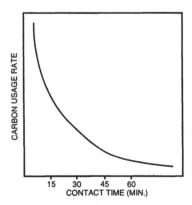

FIG. 5.2.12 Optimum carbon contact time. (Reprinted from E.K. Nyer, 1992, *Groundwater treatment technology,* 2d ed., New York: Van Nostrand Reinhold.)

suspended solids, oil and grease, and a high concentration of iron and manganese can foul the carbon and require frequent backwashing. GAC showed poor adsorption capacity for wastewaters with high fatty acids (i.e., leachate from young landfills) or wastewaters with high BOD/COD and COD/TOC ratios (U.S. EPA 1977). Furthermore, the amount of carbon required, the frequency of regeneration and reactivation, and the potential need to handle the discarded carbon as a hazardous waste make GAC a relatively expensive technology.

AIR STRIPPING

Air stripping is a mass-transfer process whereby volatile contaminants are stripped out of the aqueous solution and into the air. The process exposes the contaminated water to a fresh air supply which results in a net mass transfer of contaminants from the liquid phase to the gaseous phase. Contaminants are not destroyed by air stripping but rather are transferred into the air stream where they may need further treatment. Air stripping applies to volatile and semivolatile organic compounds. It does not apply to low volatility compounds, metals, or inorganic contaminants.

Several types of air stripping technologies are available including tray aeration, spray aeration, and packed towers. Among these technologies, packed tower aeration (PTA) is the most commonly applied to remove volatile organics from groundwater. In a packed tower, the contaminated water comes in contact with a countercurrent flow of air. The packing material in the tower breaks the water into small droplets and thin films causing a large contact area where the mass transfer can take place. Figure 5.2.14 shows a typical treatment process using air stripping.

TABLE 5.2.4 CARBON ADSORPTION WITH PPM INFLUENT LEVELS

System No.	Contaminants	Typical Influent Conc. (mg/liter)	Typical Effluent Conc. (mg/liter)	Surface Loading (gpm/ft²)	Total Contact Time (min)	Carbon Usage Rate (lb/1000 gal)	Operating Mode
1	Phenol	63	<1	1.0	201	5.8	Three fixed beds in series
	Orthorchlorophenol	100	<1				
2	Chloroform	3.4	<1	0.5	262	11.6	Two fixed beds in series
	Carbon tetrachloride	135	<1				
	Tetrachloroethylene	3	<1				
	Tetrachloroethylene	70	<1				
3	Chloroform	0.8	<1	2.3	58	2.8	Two fixed beds in series
	Carbon tetrachloride	10.0	<1				
	Tetrachloroethylene	15.0	<1				
4	Benzene	0.4	<1	1.21	112	1.9	Two fixed beds in series
	Tetrachloroethylene	4.5	<1				
5	Chloroform	1.4	<1	1.6	41	1.15	Two fixed beds in series
	Carbon tetrachloride	1.0	<1				
6	Trichloroethylene	3.8	<1	2.4	36	1.54	Two fixed beds in series
	Xylene	0.2–0.5	<1				
	Isopropyl Alcohol	0.2	<10				
	Acetone	0.1	<10				
7	Di-isoproply methyl phosphonate	1.25	<50	2.2	30	0.7	Single fixed bed
	Dichloropentadiene	0.45	<10				
1	1,1,1-Trichloroethane	143	<1	4.5	15	0.4	Single fixed bed in series
	Trichloroethylene	8.4	<1				
	Tetrachloothylene	26	<1				
2	Methyl T-butyl ether	30	<5	5.7	12	0.62	Two single fixed beds
	Di-isopropyl ether	35	<1				
3	Chloroform	400	<100	2.5	26	1.19	Four single fixed beds
	Trichloroethylene	10	<1				
4	Trichloroethylene	35	<1	3.3	21	0.21	Three single fixed series
	Tetrachloroethylene	170	<1				
5	1,1,1-Trichlorethane	70	<1	4.5	30	0.45	Two fixed beds in series
	1,1-Dichloroethylene	10	<1				
6	Trichlorethylene	25	<1	2.0	35	0.32	Single fixed bed
	Cis-1,2-dicloroethylene	15	<1				
7	Trichlorethylene	50	<1	1.6	42	0.38	Two single fixed beds
8	Cis-1,2-dichloroethylene	5	<1	1.91	70	0.25	Two fixed beds in series
	Trichloroethylene	5	<1				
	Tetrachloroethylene	10	<1				

Source: R.P. O'Brien, 1983, There is an answer to groundwater contamination, *Water/Engineering and Management* (May).

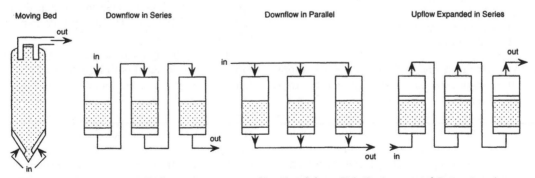

FIG. 5.2.13 Single- and multiple-stage contactors. (Reprinted from U.S. Environmental Protection Agency, 1985, *Handbook, remedial action at waste disposal sites,* EPA/625/6-85/006, Washington, D.C.: U.S. EPA.)

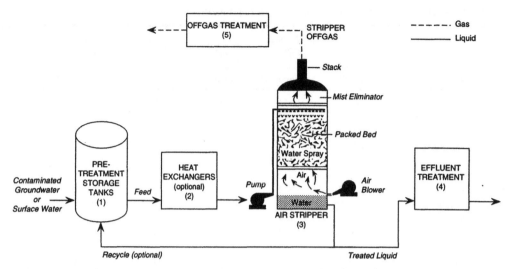

FIG. 5.2.14 Typical treatment process using air stripping. (Reprinted from U.S. Environmental Protection Agency, 1991, Air stripping of aqueous solutions, *Engineering Bulletin*, EPA/540/2-91/022, Washington, D.C.: U.S. EPA.)

Design Consideration

The design of an air stripper is based on the flow rate, type of contaminant, concentration, temperature, and effluent requirements. The major design variables are the type of packing, gas pressure drop, and air-to-water ratio. Given those design variables, environmental engineers can determine the gas and liquid loading rates, tower diameter, and packing height by using the following mass-balance equation (Noonan and Curtis 1990):

$$Z_t = \frac{L}{K_1a}\left[\frac{R}{R-1}\right] \cdot \ln \frac{\frac{C_i}{C_e}(R-1)+1}{R} \qquad 5.2(10)$$

$$D = \left[\frac{4Q}{\pi L}\right]^{0.5} \qquad 5.2(11)$$

where:

Z_t = depth of packing, m
D = diameter of the tower, m
L = liquid loading rate, m³/m²/sec
K_1a = overall liquid mass-transfer coefficient, sec⁻¹
R = stripping factor, dimensionless
C_i = influent concentration, mg/l
C_e = effluent concentration, mg/l
Q = flow rate m³/sec

The key variables to define in the preceding equations are the overall mass-transfer coefficient K_1a and the stripping factor R. The mass-transfer coefficient is a function of the type of packing, the liquid and gas flow rates, and the viscosity and density of the water. Therefore, the mass-transfer coefficient is usually determined from a pilot test on actual field data. When pilot testing is not feasible, theoretical correlations, such as those developed by Onda,

Takeuchi, and Okumoto (1968), can be used.

The stripping factor R is related to the air–water ratio as follows (Noonan and Curtis 1990):

$$R = \frac{(G/L)_{act}}{(G/L)_{min}} \qquad 5.2(12)$$

$$(G/L)_{min} = \frac{1}{H}\frac{C_i - C_e}{C_i} \qquad 5.2(13)$$

where:

$(G/L)_{min}$ = minimum air–water ratio, dimensionless
$(G/L)_{actual}$ = actual air–water ratio, dimensionless
G = gas (air) loading rate, m³/m²/sec
L = liquid loading rate, m³/m²/sec
H = Henry's constant, dimensionless

The actual air–water ratio, however, is related to the gas pressure drop through the column as shown in Figure 5.2.15 (brand-specific pressure drop curves are available from packing vendors). Therefore, engineers should examine several combinations of air–water ratio and pressure drop to determine the most cost-effective design. A high pressure drop reduces the size of the tower and capital costs; however, it increases the size of the blower and operation costs. Studies have shown that the most cost-effective stripping factor R usually falls between 3 and 5 (Hand et al. 1986).

After a stripping factor is selected, the actual air–water ratio can be calculated with Equation 5.2(12), and the gas (air) loading rate can be obtained from Figure 5.2.15 for a given pressure drop. Then the tower height and diameter can be calculated with Equations 5.2(10) and 5.2(11), respectively. This procedure should be repeated for several combinations of stripping factor and pressure drop until the most cost-effective design is obtained. Several computer cost models can be used in this process (Nirmalakhandan,

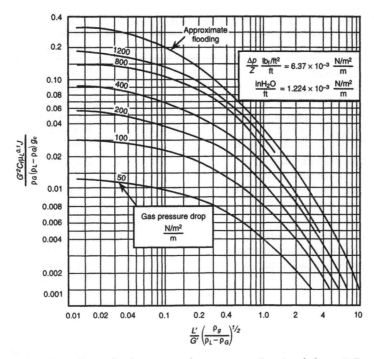

FIG. 5.2.15 Generalized pressure drop curves. (Reprinted from R.E. Treybal, 1980, *Mass transfer operations,* 3d ed., New York: McGraw-Hill.)

Lee, and Speece 1987; Cummins and Westrick 1983; Clark, Eilers, and Goodrick 1984).

Methods of Construction

The components of a stripping tower include the tower shell, tower internals, packing, and air delivery systems. The tower shell can be made of aluminum, fiberglass, stainless steel, or coated carbon steel. Selecting the shell construction material is usually based on cost, structural strength, resistance to corrosion, and esthetics. Table 5.2.5 shows the advantages and disadvantages of several materials of construction.

Tower internals include the water distributor system inside the tower, the mist eliminator system, and the air exhaust ports. The environmental engineer should select the type of components that ensure optimal mass-transfer conditions at the most economical cost.

The packing material is an important component of the air stripping tower. Several types of packing materials are commercially available including plastic, metal, or ceramic (Perry and Green 1984). The selection is based on materials exhibiting a high mass-transfer rate and a low gas pressure drop. Plastic packings are often used because of their low price, corrosion resistance, and light weight. Table 5.2.6 shows the physical characteristics of common packing materials.

Other components of an air stripping system include the blower, noise control devices, and air filters. The blower is designed based on the air–water ratio and can be mounted on top of the tower or at the base. Sound mufflers control noise, and air filters prevent contact between the water and the air outside the tower.

Operation and Maintenance

In a packed air stripping tower, the water flows countercurrent to the air stream which is introduced at the bottom of the tower. In some configurations (e.g., induced draft systems), the air is drawn through the tower by the blower instead of being forced. The water supply pumps usually control the blowers to coordinate the air and water flows. The offgas from an air stripper may need to be treated, depending on air emission requirements, with the use of granular activated carbon, catalytic oxidation, or incineration (U.S. EPA 1985a). The liquid effluent from the air stripper may contain trace amounts of contaminants which can be treated by GAC.

Maintenance of air stripping systems is minimal and usually involves the blower. However, periodic inspection of the packing is required if the water contains high levels of iron, suspended solids, or microbial population. During the aeration process, dissolved iron and manganese can be oxidized and deposited on the packing material. This deposit can build up and clog the packed bed and, therefore, reduce system efficiency. Pretreating the influent can control iron deposition. A high microbial population can lead to a biological build up within the packed

TABLE 5.2.5   CONSTRUCTION MATERIALS FOR TOWER SHELLS WITH A PACKED AIR STRIPPER

Material	Advantages	Disadvantages
Aluminum	Lightweight Low cost Corrosion resistant Excellent structural properties Long life (> fifteen years) No special coating required	Poor resistance to water with pH less than 4.5 and greater than 8.6 Pitting corrosion occurs in the presence of heavy metals Not well suited to high chloride water
Carbon Steel	Mid-range capital cost Good structural properties Long life if properly painted and maintained	Requires coating inside and outside to prevent corrosion, leading to increased maintenance Heavier than aluminum or FRP
Fiberglass	Low cost High chemical resistance to acidic and basic conditions, chlorides, and metals	Poorly defined structural properties Short life (< ten years) unless more expensive resins used Poor resistance to ultraviolet (UV) light (can be overcome with special coatings that must be maintained) Requires guy wires in most situations Susceptible to extremes of temperature differential disturbing tower shape and interfering with distribution
Stainless Steel	Highly corrosion resistant Excellent structural properties Long life (> twenty years) No special coating required	Most expensive material for prefabricated towers Susceptible to stress fracture corrosion in the presence of high chloride levels
Concrete	Aesthetics Less prone to vandalism	Difficult to cast in one place leading to potential difficulties with cracks and leaks More expensive than self-supporting prefab towers
Metal lined block and brick	Aesthetics Less prone to vandalism Prefab air stripper insert eliminates problems associated with cast in place towers	More expensive than self-supporting prefab towers

Source: K.E. Nyer, 1992, *Groundwater treatment technology*, 2d ed., New York: Van Nostrand Reinhold.

bed and reduce system performance. This problem can also be prevented through pretreatment of the influent.

Cost

The capital costs of an air stripper include the costs of the tower shell, packing, tower internals, air delivery system, electrical equipment and controls, housing, design, and contingencies. The addition of an air treatment system roughly doubles the cost of an air stripping system (Lenzo and Sullivan 1989; U.S. EPA 1986a). The cost depends on the flow rate, volatility of the contaminant, concentration, and removal efficiency. Costs vary from $0.07–0.70/1000 gal for Henry's law coefficients of 0.01–1.0 to $7.00/1000 gal for Henry's law coefficients lower than 0.005 (Adams and Clark 1991).

Advantages and Limitations

Air stripping is a proven technology for treating water contaminated with volatile and semivolatile organic compounds. Removal efficiencies of greater than 98% for volatile organics and greater than or equal to 80% for semivolatile compounds have been achieved. Recent developments in this technology include high temperature air stripping and air rotary stripping to increase removal efficiencies (Bass and Sylvia 1992). The use of diffused air or bubble aeration air strippers for flows less than 50 gpm have also increased during the last five years.

The air stripping technology, however, is not effective in treating low volatility compounds, metals, or inorganics. Air emissions of volatile organics from the air stripper may need a separate treatment. In addition, the removal efficiency of air strippers is reduced for aqueous solutions

TABLE 5.2.6 PHYSICAL CHARACTERISTICS OF COMMON PACKING MATERIALS

Type	Size	Surface Area (ft^3/ft^3)	Void Space (%)	Packing Factor[a]
Dumped Packings				
Glitsch	0A	106	89	60
Mini-rings	1A	60.3	92	30
(Plastic)	1	44	94	28
	2A	41	94	28
	2	29.5	95	15
	3A	24	95.5	12
Tellerettes	1″(#1)	55	87	40
(Plastic)	2″(2-R)	38	93	18
	3″(3-R)	30	92	16
	3″(2-K)	28	95	12
Intalox	1″	63	91	33
Saddles	2″	33	93	21
(Plastic)	3″	27	94	16
Pall rings	⅝″	104	87	97
(Plastic)	1″	63	90	52
	1½″	39	91	40
	2½″	31	92	25
	3½″	26	92	16
Raschig rings	½″	111	63	580
(Ceramic)	¾″	80	63	255
	1″	58	73	155
	1½″	38	71	95
	2″	28	74	65
	3″	19	78	37
Jaegar	1″	85	90	28
Tri-Packs	2″	48	93	16
(Plastic)	3½″	38	95	12
Stacked Packing				
Delta	—	90	98	—
(PVC)				
Flexipac	Type 1	170	91	33
(Plastic)	Type 2	75	93	22
	Type 3	41	96	16
	Type 4	21	98	9

Source: R.E. Treybal, 1980, *Mass transfer operations,* 3d ed., New York: McGraw-Hill.

with high levels of suspended solids, iron, manganese, or microbial population. Periodic cleaning of the packing material removes the deposits of these products.

OXIDATION AND REDUCTION

In chemical oxidation, the oxidation state of a contaminant is increased by the loss of electrons, while the oxidation state of the reactant is lowered. Conversely, in reduction, the oxidation state of a contaminant is decreased by the addition of electrons. Oxidizing or reducing agents can be added to contaminated water to destroy, detoxify, or convert the contaminants to less hazardous compounds.

Many hazardous substances including various organics, sulfites, soluble cyanide- and arsenic-containing compounds, hydroxylamine, and chromates can be oxidized or reduced to forms which are more readily removed from groundwater (Huibregts and Kastman 1979).

Chemical Oxidation

Chemical oxidation involves adding oxidizing agents to the contaminated water and maintaining the pH at a proper level. The choice of an oxidizing agent depends on the substance or substances to be detoxified. Numerous oxidizing agents are available to detoxify a variety of compounds. The most commonly used agents are hydrogen peroxide, ozone, hypochlorite, chlorine, and chlorine dioxide because they tend not to form toxic compounds or residuals and are relatively inexpensive. Ozone and hydrogen peroxide have an advantage over oxidants containing chlorine because potentially hazardous chlorinated compounds are not formed (U.S. EPA 1986b).

Hydrogen peroxide is a stable and readily available substance that can oxidize many compounds. Industrial treatment plants have used hydrogen peroxide to detoxify cyanide and organic pollutants including formaldehyde, phenol, acetic acid, lignin sugars, surfactants, amines and glycol ethers, aldehydes, dialkyl sulfides, dithionate, and certain nitrogen and sulfur compounds (Envirosphere Company 1983).

Ozone is a strong oxidizing agent (gas) that is unstable and extremely reactive. Therefore, ozone cannot be shipped or stored but must be generated onsite immediately prior to application (U.S. EPA 1985b). Ozone rapidly decomposes to oxygen in solutions containing impurities. Ozone's half-life in distilled water at 68°F is twenty-five minutes, while in groundwater it drops to eighteen minutes (Envirosphere Company 1983).

Hypochlorite is used in drinking water and municipal wastewater systems for the treatment and control of algae and biofouling organisms (U.S. EPA 1985b). In industrial waste treatments, hypochlorite is used for the oxidation of cyanide, ammonium sulfide, and ammonium sulfite (Huibregts and Kastman 1979). Sodium hypochlorite solutions at concentrations of 2500 mg/l are also used for the detoxification (by oxidation) of cyanide contamination from indiscriminate dumping (Farb 1978). However, because the principal products from chlorination of organic contaminants are chlorinated organics which can be as much of a problem as the original compound, hypochlorite treatment is limited.

Advanced Oxidation

Advanced oxidation uses UV radiation combined with ozone or hydrogen peroxide to enhance the oxidation rate of the compounds; reaction times can be 100 to 1000 times faster in the presence of UV light (U.S. EPA 1986b). UV light reacts with hydrogen peroxide molecules to form an hydroxyl radical, a powerful chemical oxidant. Specifically, hydrogen peroxide and UV light are used as shown in Figure 5.2.16 for the treatment of volatile organic compounds and other organic contaminants in contaminated groundwater (U.S. EPA 1993). In addition, hydrogen peroxide, ozone, and UV radiation are used as shown in Figure 5.2.17 for the oxidation of dissolved organic contaminants including chlorinated hydrocarbons and aromatic compounds in groundwater (U.S. EPA 1990).

Chemical Reduction

Environmental engineers have proposed chemical reduction to detoxify wastes and contaminated waters, but its application does not appear to have the potential that chemical oxidation has. For example, they have proposed sodium sulfites to treat groundwater contaminated by sodium hypochlorite (Huibregts and Kastman 1979) and ferrous sulfate in conjunction with hydroxides to detoxify and insolubilize hexavalent chromium (Tolman et al. 1978; Metcalf and Eddy, Inc. 1972). Little work has been done in the use of chemical reduction for organic wastes.

Cost

Costs for oxidation systems include the costs for storage and handling equipment, chemicals, feed systems and controls, and electricity to operate the ozone generator or the UV lamps. Costs for enhanced oxidation range from $0.15 to $70/1000 gal treated depending on the type of contaminants, their concentration, and the cleanup level (U.S. EPA 1993, 1990).

Advantages and Limitations

The principal advantage of chemical oxidation technology is the ability of oxidizing agents to degrade carbonaceous compounds, theoretically to carbon dioxide and water (Roy 1990b). Adequate oxidant and operating conditions (i.e., temperature, pH, and contact time), however, must be present to facilitate a complete reaction. Incomplete reactions can generate partially oxidized products which may require further treatment. Oil and grease in the water can minimize the efficiency of the oxidation process. In addition, UV lamps do not perform well in turbid waters because of the reduced light transmission (Roy 1990a).

Limitations of Pump-and-Treat Technologies

Pump-and-treat is the most commonly used technology for groundwater remediation and plume containment. However, recently pump-and-treat technology has been subject to increasing scrutiny and controversy. One significant problem with the technology is its inability to achieve cleanup goals within reasonable time frames (Galya 1994). At many sites where this technology is used,

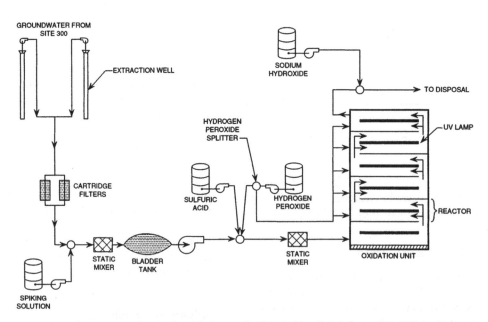

FIG. 5.2.16 Perox-pure chemical oxidation technology. (Reprinted from U.S. Environmental Protection Agency, 1993, *Perox-pure chemical oxidation technology—Perioxidation Systems, Inc.,* EPA/540/AR-93/501 Superfund Innovative Technology Evaluation, Washington, D.C.: U.S. EPA.)

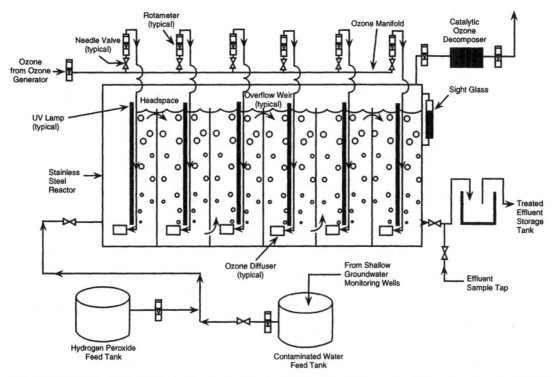

FIG. 5.2.17 Ultrox, UV/oxidation technology. (Reprinted from U.S. Environmental Protection Agency, 1990, *Ultraviolet radiation/oxidation technology—Ultrox International,* EPA/540/AS-89/012 Superfund Innovative Technology Evaluation, Washington, D.C.: U.S. EPA.)

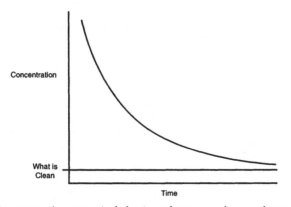

FIG. 5.2.18 Assumptotic behavior of pump-and-treat cleanup technologies. (Reprinted from K.E. Nyer, 1992, *Groundwater treatment technology,* 2d ed., New York: Van Nostrand Reinhold.)

contaminant removal rates follow a relatively consistent pattern. After a period of initially steady reductions, groundwater contaminant concentrations tend to level off and remain fairly constant, with random fluctuations around an assumptotic limit (Tucker et al. 1989) as shown in Figure 5.2.18. The assumptotic concentration level may be higher than the specified cleanup target, and achieving cleanup goals within reasonable time frames may not be possible.

Therefore, pump-and-treat technology is not an effective approach by itself for the ultimate remediation of aquifers to health-based cleanup concentrations.

—*Ahmed Hamidi*

References

Adams, C.E., and W.W. Eckenfelder. 1974. *Process design techniques for industrial waste treatment.* Nashville, Tenn.: Enviro Press.

Adams, J.Q., and R.M. Clark. 1991. Evaluating the cost of packed-tower aeration and GAC for controlling selected organics. *Journal AWWA* 1:49–57.

Bass, D.H., and T.E. Sylvia. 1992. Heated air stripping for the removal of MTBE from recovered groundwater. Proceedings for the 1992 Petroleum Hydrocarbons and Organic Chemicals in Groundwater, 4–6 November, Houston, Tex.

Campbell, M.D., and J.H. Lehr. 1977. Well cost analysis. In *Water well technology.* 4th ed. McGraw-Hill.

Clark, R.M., R.G. Eilers, and J.A. Goodrick. 1984. VOCs in drinking water: Cost of removal. *Journal of Environmental Engineering* 110, no. 6:1146–1162.

Cohen, R.M., and W.J. Miller. 1983. Use of analytical models for evaluating corrective actions at hazardous waste disposal facilities. Proceedings of the Third National Symposium on Aquifer Restoration and Groundwater Monitoring. Worthington, Ohio: National Water Well Association.

Corbitt, R.A. 1990. Wastewater disposal. In *Standard handbook of environmental engineering,* edited by R.A. Corbitt. New York: McGraw-Hill.

Crittenden, J.C. et al. 1988. Using GAC to remove VOCs from air stripper off-gas. *Journal AWWA* 80, no. 5(May):73–84.

Cummins, M.D., and J.J. Westrick. 1983. Trichloroethylene removal by packed column air stripping: Field verified design procedure. In Proceedings ASCE Environmental Engineering Conference, 442–449. Boulder, Colo.: ASCE.

Envirosphere Company. 1983. *Evaluation of systems to accelerate stabilization of waste piles or deposits.* Cincinnati, Ohio: U.S. EPA.

Farb, D. 1978. *Upgrading hazardous waste disposal sites: Remedial approaches.* EPA-SW-677. Cincinnati, Ohio: U.S. EPA.

Galya, D. 1994. Evaluation of the effectiveness of pump and treat groundwater remediation system. In WEF Specialty Conference Series Proceedings on Innovative Solutions for Contaminated Site Management, 6–9 March, 323–342. WEF.

Hand, D.W. et al. 1986. Design and evaluation of an air stripping tower for removing VOCs from groundwater. *Journal of AWWA* 78, no. 9:87–97.

Huibregts, K.R., and K.H. Kastman. 1979. Development of a system to protect groundwater threatened by hazardous spills on land. Oil and Hazardous Material Spills Brands, Industrial Environmental Research Laboratory. Edison, N.J.: U.S. EPA.

Johnson Division, UOP, Inc. 1975. *Groundwater and wells.* Saint Paul, Minn.: Edward F. Johnson, Inc.

Kuffs, C. et al. 1983. Procedures and techniques for controlling the migration of leachate plumes. Ninth Annual Research Symposium on Land Disposal, Incineration and Treatment of Hazardous Waste.

Lenzo, F., and K. Sullivan. 1989. Groundwater treatment techniques, an overview of state-of-the-art in America. First US/USSR Conference on Hydrogeology, July. Moscow.

Luthin, J.N. 1957. *Drainage of agricultural lands.* Madison, Wis.: American Society of Agronomy.

Means. 1994. *Means site work and landscape cost data.* Means Southam Construction Information Network.

Metcalf and Eddy, Inc. 1972. *Wastewater engineering: Collection, treatment, and disposal.* New York: McGraw-Hill.

Nirmalakhandan, N., Y.H. Lee, and R.E. Speece. 1987. Designing a cost effective air stripping process. *Journal of AWWA* 79, no. 1:56–63.

Noonan, D.C., and J.T. Curtis. 1990. *Groundwater remediation and petroleum: A guide for underground storage tanks.* Chelsea, Mich.: Lewis Publishers.

Oakley, D. et al. 1994. The use of horizontal wells in remediating and containing a jet fuel plume—preliminary findings. WEF Specialty Conference Series Proceedings on Innovative Solutions for Contaminated Site Management, March, 331–342. Water Environment Federation.

O'Brien, R.P. 1983. There is an answer to groundwater contamination. *Water/Engineering and Management* (May).

Onda, K.H., Takeuchi, and Y. Okumoto. 1968. Mass transfer coefficients between gas and liquid phases in packed columns. *Journal of Chemical Engineering*, Japan 72, no. 12:684.

Perry, R.H., and D. Green. 1984. *Perry's chemical engineer's handbook.*

6th ed. New York: McGraw-Hill.

Powers, J.P. 1981. *Construction dewatering: A guide to theory and practice.* New York: John Wiley and Sons.

Repa, E. et al. 1982. *The establishment of guidelines for modeling groundwater contamination from hazardous waste facilities.* JRB Assoc. report prepared for the Office of Solid Waste, U.S. EPA.

Roy, K. 1990a. Researchers use UV light for VOC destruction. *Hazmat World* (May):82–93.

———. 1990b. UV-oxidation technology, shining star or flash in the pan? *Hazmat World* (June):35–50.

Soil Conservation Service. 1973. *Drainage of agricultural land.* Syosset, N.Y.: Water Information Center.

Tolman, A. et al. 1978. *Guidance manual for minimizing pollution from waste disposal sites.* EPA/600/2-78/142. Cincinnati, Ohio: U.S. EPA.

Tucker, W.A. et al. 1989. Technological limits of groundwater remediation: A statistical evaluation method. Proceedings of the Petroleum Hydrocarbons and Organic Chemicals in Groundwater, 15–17 Nov., Houston, Tex.

U.S. Environmental Protection Agency. 1977. *Wastewater treatment facilities for sewered small communities.* Washington, D.C.: U.S. EPA, Technology Transfer Division.

———. 1985a. *Handbook, remedial action at waste disposal sites.* EPA/625/6-85/006. Washington, D.C.: U.S. EPA.

———. 1985b. *Leachate plume management.* EPA/540/2-85/004. Washington, D.C.: U.S. EPA.

———. 1986a. *Mobile treatment technologies for superfund wastes.* EPA/540/2-86/003(f). Washington, D.C.: U.S. EPA.

———. 1986b. *Systems to accelerate in situ stabilization of waste deposits.* EPA/540/2-86/002. Cincinnati, Ohio: U.S. EPA.

———. 1988. A compendium of technologies used in the treatment of hazardous waste. EPA/540/2-88/1004. Washington, D.C.: U.S. EPA.

———. 1989. *Seminar publication on transport and fate of contaminants in the subsurface.* EPA/625/4-89/019. Cincinnati, Ohio: U.S. EPA.

———. 1990. *Ultraviolet radiation/oxidation technology—Ultrox International.* EPA/540/A5-89/012, Superfund Innovative Technology Evaluation. Washington, D.C.: U.S. EPA.

———. 1993. *Perox-pure chemical oxidation technology—Perioxidation Systems, Inc.* EPA/540/AR-93/501 Superfund Innovative Technology Evaluation. Washington, D.C.: U.S. EPA.

Van Hoorn, J.W., and W.H. Vandemolen. 1974. *Drainage of slopping of lands, drainage principles and applications.* Vol. 4 of *Design and management of drainage systems.* Publ. 16. 329–339. Wageningen, The Netherlands: International Institute of Land Reclamation Improvements.

Wasseling, J. 1973. Theories of field drainage and watershed runoff: Subsurface flow into drains. In *Drainage, principles and applications.* Wageningen, The Netherlands: International Institute for Land Reclamation and Improvement.

5.3
IN SITU TREATMENT TECHNOLOGIES

In situ treatment is an alternative to pump-and-treat technology and involves the underground destruction and neutralization of contaminants. The technology has the advantage of requiring minimal surface facilities and reducing public exposure to the contaminant. Theoretically, the technology could be applied to both organic and inorganic contaminants. However, in situ treatment is still relatively new and for the most part has been limited to organic compounds. The most commonly used in situ treatment methods include bioremediation, air sparging, and chemical detoxification.

Bioremediation

Bioremediation is a relatively new technology that has recently gained considerable attention. Bioremediation uses naturally occurring microorganisms to degrade and break down organic contaminants into harmless products consisting mainly of carbon dioxide and water. In situ bioremediation has two basic approaches. The first approach relies on the natural biological activities of indigenous microorganisms in the subsurface. The second approach is called *enhanced bioremediation* and involves stimulating the existing microorganisms by adding oxygen and nutrients. Most organic compounds are biodegradable, some faster than others. The rate of biodegradation, however, depends on the chemical structure of the compound as discussed in Section 3.2 and shown in Table 3.3.2. Figure 5.3.1 shows a simplified representation of a groundwater bioremediation system.

DESIGN CONSIDERATIONS

The design variables of bioremediation include the amount of bacteria, oxygen, and nutrients needed for the biodegradability of the contaminant as well as the characteristics of the subsurface environment. Given those variables, environmental engineers can determine an appropriate hydraulic design of the bioremediation system. Computer models such as BIOPLUME II (1986) can assist in the design of bioremediation systems.

The number of bacteria must be sufficient to consume all of the organic contaminants in a timely manner. Most sites have significant populations of indigenous microorganisms that can degrade a variety of organic contaminants. One gram of surface soil can contain from 0.1 to 1 billion cells of bacteria, 10 to 100 million cells of actinomycetes, and 0.1 to 1 million cells of fungi (Dockins 1980;

Whitelaw and Edwards 1980). The microorganism population in soils is generally greatest in the surface horizons where the temperature, moisture, and energy supply is favorable for their growth. As the depth increases, the number of aerobic microorganisms decreases; however, anaerobic microorganisms can exist depending on the availability of nutrients and organic material. The type of microorganisms present on site and their optimal living conditions can be determined in the laboratory. If indigenous microorganisms are not present on site or if their number is not sufficient to consume all organic contaminants, appropriate exogenous microorganisms can be imported, or existing microorganisms can be stimulated with the addition of oxygen and nutrients.

In addition, aerobic bacteria require oxygen for their growth. Because the concentrations of dissolved oxygen in groundwater are generally low, adding oxygen supports the aerobic biodegradation of organic compounds in groundwater. The theoretical quantities of oxygen required to degrade an organic compound can be determined from stoichiometric analysis. For example, degradation of a simple organic acid, such as acetic acid, theoretically requires 1.1 mg of oxygen. Oxygen can be added in several ways, including aeration, oxygenation, and the use of hydrogen peroxide and other oxygen-containing compounds. Obviously, the use of these compounds requires careful control of the geochemistry and hydrology of the site.

Inorganic nutrients including nitrogen, phosphorous, and potassium are needed for proper bacterial growth and can limit cell growth if they are not present at sufficient levels. The groundwater may already contain levels of phosphorous and nitrogen, but these levels are probably insufficient for bacterial growth (Bouwer 1978; Doetsch and Cook 1973). The addition of nutrients, however, can contaminate the aquifer. Therefore, only the amount needed to sustain biological activity should be added.

Other factors limit the growth rate of bacteria and, therefore, the biodegradation of organic contaminants in groundwater. These factors include the pH, temperature, and toxicity of the contaminant. The appropriate range for these parameters should be determined in a treatability study.

ADVANTAGES AND LIMITATIONS

Bioremediation has several advantages over other cleanup technologies including cost, minimal surface facilities, and minimum public exposure to the contaminant. However,

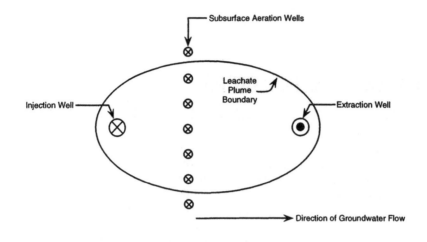

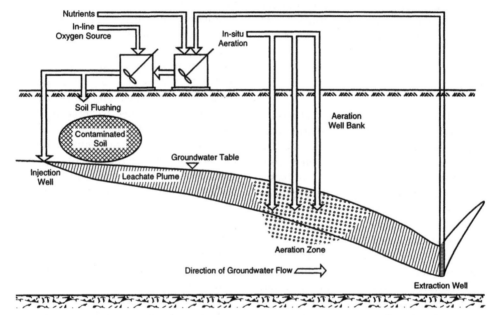

FIG. 5.3.1 Simplified representation of a groundwater bioremediation system. (Reprinted from U.S. Environmental Protection Agency, 1985, *Handbook, remedial action at waste disposal sites,* EPA/625/6-85/006, Washington, D.C.: U.S. EPA.)

bioremediation suffers from several drawbacks (Lee et al. 1988). The technology is limited to aquifers with high permeability. Bacterial growth can be inhibited by one or more compounds at sites with mixed wastes. In addition, incomplete degradation of some substances can lead to other types of contamination.

Air Sparging

Air sparging, also called *in situ stripping,* is an innovative technology that injects air into the saturated zone to remove contaminants from the water. The air injected in the saturated area creates bubbles that rise and carry trapped and dissolved contaminants into the unsaturated zone above the water table (Camp Dresser & McKee, Inc. 1992). This technology is typically used in conjunction with soil vapor extraction (SVE) to enhance the removal rate of contaminants from the saturated and unsaturated zones (Bohler et al. 1990). As volatile organic compounds reach the unsaturated zone, they are captured by the SVE vapor wells that are screened in the unsaturated zone, as illustrated in Figure 5.3.2. Air sparging also provides an oxygen source which may stimulate bioremediation of some contaminants. Air sparging is applicable for contaminants which have a high Henry's constant or high vapor pressure in soils with high permeability.

DESIGN CONSIDERATIONS

The design variables for an air sparging system include the volatility and concentration of the compound, the porosity and permeability of the soil, and the temperature of the

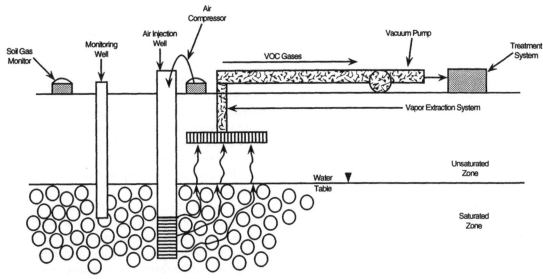

FIG. 5.3.2 Simplified representation of an air sparging system.

water. Given those variables, environmental engineers can determine the radius of influence of the air sparge wells, the air flow rate, and the vacuum pressure needed. Although the technology has been used at several sites (Loden and Fan 1992), references to the design of an air sparging system are limited (Sellets and Schreiber 1992; Marley, Li, and Magee 1992), and in most cases the design is based on empirical formulas or the results of pilot studies.

ADVANTAGES AND LIMITATIONS

Air sparging is a promising technology which has several advantages. Air sparging can extend the effectiveness of SVE systems to include volatile contaminants from the saturated zone, and the contaminants can be treated onsite without removal or potential public exposure to them. In addition, air sparging can expedite groundwater cleanup. The technology, however, is limited to aquifers with high permeability and contaminants with high volatility. In addition, the technology is relatively new, and the number of case studies where the technology has been successfully applied is limited.

Other Innovative Technologies

Over the last few years, several innovative technologies have been proposed for the in situ treatment of groundwater. Although these technologies have not yet been developed to the extent of previously discussed technologies, some of them have demonstrated success in actual site remediations (Wagner et al. 1986). Laboratory and pilot testing, however, are necessary to evaluate the applicability of

a particular technology to a site. Examples of innovative groundwater technologies are described next.

NEUTRALIZATION AND DETOXIFICATION

In situ neutralization and detoxification involves injecting a substance into groundwater that neutralizes or destroys a contaminant. The technology is limited to contaminants that can be neutralized or degraded to nontoxic byproducts. Neutralization and detoxification is applicable to both organic and inorganic compounds. Selecting a treatment agent depends on the type of contaminant and the characteristics of the subsurface environment such as temperature, permeability, pH, salinity, and conductivity. Examples of in situ treatment agents include hydrogen peroxide that can be injected directly into groundwater through existing monitoring wells or subsurface drains (Vigneri 1994). Hydrogen peroxide produces the hydrogen free radical OH, an extremely powerful oxidizer which progressively reacts with organic contaminants to produce carbon dioxide and water.

Other in situ neutralization and detoxification technologies include precipitation and polymerization. Precipitation involves injecting substances into the groundwater plume which form insoluble products with the contaminants, thereby reducing the potential for migration in groundwater (U.S. EPA 1985). This technique is mainly applicable to dissolved metals, such as lead, cadmium, zinc, and iron. Some forms of arsenic, chromium, and mercury and some organic fatty acids can also be treated by precipitation (Huibregts and Kastman 1979). The most common precipitation reagents include hydroxides, oxides, sulfides, and sulfates. As with other in situ techniques, precipitation is only applicable to sites with aquifers having high hydraulic conductivities. The major disadvantages of precipitation are that it can only be applied to a narrow, specific group of chemicals (mainly metals); that a potential groundwater pollutant may be injected; that toxic gases (as in sulfide treatment) may form; and that the precipate may resolubilize (U.S. EPA 1985).

In situ polymerization involves injecting a polymerization catalyst into the nonaqueous organic phase of a contaminant plume to cause polymerization (U.S. EPA 1985). The resulting polymer is gel-like and nonmobile in the groundwater flow regime. Polymerization is a specific technique that is applicable to organic monomers such as styrene, vinyl chloride isoprene, methyl methacrylate, and acrylonitrile (Huibregts and Kastman 1979). In a hazardous waste site where groundwater pollution has occurred over time, any organic monomers originally present would most likely have polymerized upon contact with the soil (U.S. EPA 1985). Therefore, in situ polymerization is a technique most suited for groundwater cleanup following land spills or underground leaks of a pure monomer. The major disadvantages of polymerization include its limited application and the difficulty of initiating sufficient contact of the catalyst with the dispersed monomer (Huibregts and Kastman 1979).

PERMEABLE TREATMENT BEDS

Permeable treatment beds are also in situ treatment techniques used at sites with relatively shallow groundwater tables. The concept of a permeable treatment bed involves excavating a trench, filling the trench with a permeable treatment material, and allowing the plume to flow through the bed thus physically removing or chemically altering the contaminants. The function of a permeable treatment bed is to reduce the quantities of contaminants in the plume to acceptable levels. Potential problems with using a permeable treatment bed include saturation of the bed material, plugging of the bed with precipitates, and the short life of the treatment material (U.S. EPA 1985).

The selection of the appropriate bed material to treat the contaminants and the design of the bed are two elements that determine the effectiveness of a permeable treatment bed. The types of available treatment bed fill material include limestone, crushed shell, activated carbon, glauconitic greensands, and synthetically produced ion exchange resins. Ensuring proper physical design of the treatment bed requires a knowledge of the hydrogeology of the site (e.g., groundwater flow rate and direction, hydraulic conductivities) and the chemical characteristics of the plume (U.S. EPA 1985).

PNEUMATIC FRACTURING

Environmental engineers use pneumatic fracturing extraction and hot gas injection to treat in situ contamination located within low permeable formations (Accutech Remedial Systems, Inc. 1994). The process has been demonstrated at numerous sites and significantly increases subsurface permeability and contaminant mass removal (U.S. EPA 1993b). The process applies controlled bursts of high pressure air into a well through a proprietary injection and monitoring system. When the down-hole pressure exceeds the pressure of the formation, channels or fractures are created propagating from the fracture well. Once the permeability of the formation is increased, engineers inject hot gas air (250 to 300°F for pilot-scale and 300 to 600°F for full-scale design) under pressure to elevate the temperature of the fracture surface and volatilize contaminants located within the formation matrix. The extracted vapors are then treated by activated carbon during low-concentration process streams or by catalytic technology during high-concentration process streams.

The technology can be applied at depths to 50 feet and has a radius of influence of as much as 40 feet from the injection point (well). Subsurface air flow has been increased 150 times compared with the site's natural permeability. The technology, however, is not applicable for treating inorganic or nonvolatile organic compounds. In addition, applying the pneumatic fracturing process may be unnecessary at a site with a high natural permeability.

THERMALLY ENHANCED RECOVERY

The in situ steam enhanced extraction process, called thermally enhanced recovery (Praxis Environmental Services Inc. 1994), removes volatile and semivolatile organic compounds from an area of contaminated soil or groundwater without excavation. The process operates through the use of wells constructed in the contaminated soil. High-quality steam is added to the soil through some wells, called *injection wells*. Other wells, known as *extraction wells*, operate under vacuum to remove liquid and vapor contaminants and water from the soil. Injecting steam into the ground raises the temperature of the soil and causes the most volatile compounds to vaporize. In addition, pressure gradient is formed between the injection and extraction wells which drives the flow of steam and vaporized contaminants towards the extraction wells (U.S. EPA 1993a). Raising the temperature of the soil matrix also assists in removing less volatile compounds by increasing their in situ vapor pressure. After the entire soil mass being treated has reached the steam temperature, as determined by soil–temperature monitors, and steam breakthrough occurs at the extraction wells, the flow of steam continues only intermittently with a constant vacuum applied to the extraction wells. The vacuum extraction removes much of the remaining contamination. As the soil in the high permeability region cools, the steam remaining in the low permeability region evaporates the contaminants.

The technology is cost-effective for large and deep areas of contamination where technologies requiring excavation are difficult or impossible. The process can be applied in sections to treat an area of any size and depth. If the site, however, contains a high concentration (>200 ppm) of heavier-than-water organics, a possibility exists that these compounds might be mobilized downward into groundwater. In addition, treatment of shallow (<10 feet) contaminated areas is less cost-effective than deeper areas compared to other technologies.

—Ahmed Hamidi

References

Accutech Remedial Systems, Inc. 1994. *Pneumatic fracturing.* Keyport, N.J.

BIOPLUME II. 1986. *Computer model of two dimensional contaminant transport under the influence of oxygen limited biodegradation in groundwater.* Houston, Tex.: National Center for Groundwater Research, Rice University.

Bohler, U. et al. 1990. Air injection and soil air extraction as a combined method for cleaning contaminated sites: Observations from test sites in sediments and solid rocks. In *Contaminated Soil '90,* edited by F. Arench et al., 1039–1044. The Netherlands: Kluwser Academic Publ.

Bouwer, H. 1978. *Groundwater hydrology.* New York: McGraw-Hill.

Camp Dresser & McKee, Inc. 1992. *A technology assessment of soil vapor extraction and air sparging.* Risk Reduction Engineering Laboratory, Office of Research and Development. Cincinnati, Ohio: U.S. EPA.

Dockins, W.S. et al. 1980. Dissimilatory bacterial sulfate reduction in Montana groundwaters. *Geomicrobiology Journal* 2, no. 1:83–98.

Doetsch, and T.M. Cook. 1973. *Introduction to bacteria and their eco-biology.* Baltimore, Md.: University Park Press.

Huibregts, K.R., and K.H. Kastman. 1979. *Development of a system to protect groundwater threatened by hazardous spills on land.* Oil and Hazardous Material Spills Brands, Industrial Environmental Research Laboratory. Edison, N.J.: U.S. EPA.

Lee, M.D. et al. 1988. Biorestoration of aquifers contaminated with organic compounds. *CRC Crit. Rev. Environ. Control* 18:29–89.

Loden, M.E., and C.Y. Fan. 1992. Air sparging technology evaluation. Proceedings of 2nd National Research and Development Conference on the Control of Hazardous Materials, 328–334. San Francisco, Calif.

Marley, M.C., F. Li, and S. Magee. 1992. The application of a 3-D model in the design of air sparging systems. Proceedings of the Petroleum Hydrocarbons and Organic Chemicals in Groundwater 4–6 Nov., 377–392. Houston, Tex.: NGWA.

Praxis Environmental Services, Inc. 1994. *Thermally enhanced recovery in situ.* San Francisco, Calif.

Sellets, K.L., and R.P. Schreiber. 1992. Air sparging model for predicting groundwater cleanup rate. Proceedings of the Petroleum Hydrocarbons and Organic Chemicals in Groundwater, 4–6 Nov., 365–376. Houston, Tex.: NGWA.

U.S. Environmental Protection Agency. 1985. *Leachate plume management.* EPA/540/2-85/004. Washington, D.C.: U.S. EPA.

———. 1993a. In-situ steam enhanced extraction process. In *Superfund Innovative Technology Evaluation Program, Technology Profiles.* 6th ed. EPA/540/R-93/526. Washington, D.C.: U.S. EPA.

———. 1993b. Pneumatic fracturing extraction. In *Superfund Innovative Technology Evaluation Program, Technology Profiles.* 6th ed. EPA/540/R-93/526. Washington, D.C.: U.S. EPA.

Vigneri, R. 1994. *Groundwater remediation primer.* Wilmington, N.C.: Cleanox Environmental Services, Inc.

Wagner, K. et al. 1986. *Remedial action technology for waste disposal sites.* 2d ed. Park Ridge, N.J.: Noyes Data Corporation.

Whitelaw, K. and R.A. Edwards. 1980. Carbohydrates in the unsaturated zone of the chalk. *England Chemical Geology* 29, no. 314:281–291.

Stormwater Pollutant Management

David H.F. Liu | Kent K. Mao

113

6
Storm Water Pollutant Management

6.1
INTEGRATED STORM WATER PROGRAM

Storm water is defined as storm water runoff, snowmelt runoff, and surface runoff and drainage. Storm water management is important in urban water systems, including water supply systems and wastewater systems. With increasing residential, commercial, and industrial development, stormwater has become an important issue.

Growing urbanization has a significant impact on the surrounding environment, creating problems such as nonpoint sources of water pollution. Because of changes in land-use patterns, pollutants in developed areas build up during dry periods and are washed off as runoff passes over land surfaces. Nonpoint sources account for about 45%, 76% and 65% of the degradation of estuaries, lakes, and rivers respectively (EPA 1989). In comparison, municipal and industrial point source discharges under National Pollution Discharge Elimination System (NPDES) control account for about 9–30% of the degradation of these water sources.

In contrast to our complex urban environment, the hydrological cycle shown in many hydrology textbooks is rather simplistic. Modification of natural drainage paths, damming of waterways, impoundment of water, reuse of stormwater, and implementation of new stormwater management processes result in highly intricate hydrological processes. The development of storm water runoff and its possible superimposition on dry weather flow in combined sewer systems are summarized in Figure 6.1.1. A detailed urban drainage subsystem is shown in Figure 6.1.2.

Integrated Management Approach

Storm water system components and functions interact with, and may also interfere with, each other. Integrated system management coordinates actions to achieve water quantity and quality control, focusing on issues such as floodplain management, erosion and sediment control, nonpoint source pollution, and preservation of wetlands and wildlife habitat. System management also facilitates cooperation among all levels of government, and helps to implement laws and regulations to control storm water pollution.

FEDERAL PROGRAMS

In the 1987 amendments to the Clean Water Act, Congress mandated development of a permit system for certain sources of storm water discharge, thus the EPA has established permit application requirements for industrial storm water discharges and municipal storm sewer system discharges. Pollutants entering storm water and surface water systems are now regulated as point sources under Section 402(p) and subject to the NPDES permit process.

The EPA also provides assistance and guidance to municipalities developing storm water management programs. Although there are several agencies with possible authority in this field, no federal agency has assumed general responsibility or control. Most actions taken to date have been local initiatives. Only the Soil Conservation Service has long-standing programs of storm water management.

However, many federal agencies are directly involved in flood hazard mitigation, flood control, and floodplain management. Although there is no federal agency directly mandated to plan and implement stormwater management programs, there are several agencies engaged in related activities.

The federal government exerts a broad influence via its many agencies. For example, in the Corps of Engineers' major structural flood control program, the federal agency consults with local agencies, but maintains field offices and staff for planning, construction, operation, and maintenance. In another approach, the Soil Conservation Service (SCS) has a nationwide network of conservation districts. The districts perform some functions autonomously, while other functions are carried out by the federal staff. In floodplain management, the Federal Emergency Management Agency (FEMA) has established fairly complete federal control, although actions affecting individuals are legally mandated by state laws and local ordinances. In this case,

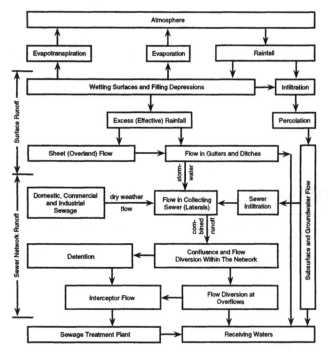

FIG. 6.1.1 Development of Stormwater Runoff and Flow in Combined Sewer Systems. (Reprinted, with permission, from W.F. Geiger, 1984, *Combined sewage quantity and quality—a contribution to urban drainage planning,* [Muenchen Technical Universitat, Muenchen].)

the financial incentives of the flood insurance program are the prime motivation for obtaining required state legislation and local ordinances.

STATE PROGRAMS

State governments enable legislation providing for involvement in storm water management. For example, the Washington State Environmental Policy Act (SEPA) ensures that environmental values are considered by state and local government officials when making decisions. The Department of Ecology (State of Washington) recently completed a storm water rule, a highway storm water rule, and a model storm water ordinance for local governments. These rulings require the development of storm water management programs for cities and counties.

MUNICIPAL PROGRAMS

County-level involvement plays an important role in implementing comprehensive storm water management plans. The principal authority for storm water management is the government with jurisdiction, usually a municipality. Municipalities usually have legal control of:

- Erosion and sedimentation ordinances
- Floodplain ordinances
- Storm water drainage ordinances
- Zoning ordinances
- Building codes
- Grading ordinances

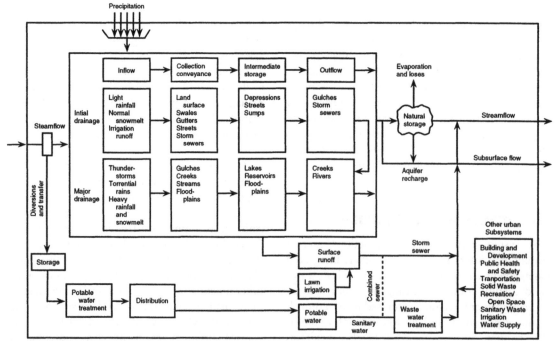

FIG. 6.1.2 The urban storm drainage subsystem.

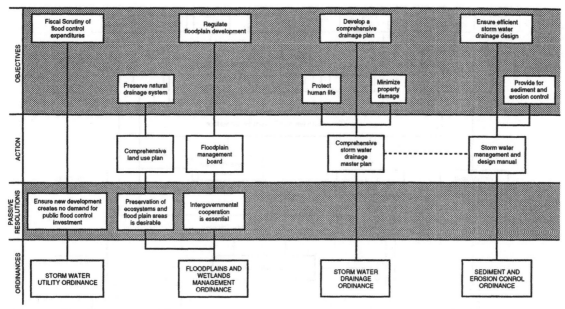

FIG. 6.1.3 Developing a comprehensive storm management program.

Storm water management is closely tied to future land use development and management. Existing and future land use development are incorporated into an integrated storm water management program as presented in Figure 6.1.3.

Many municipalities now require developers to consider future development of watersheds when designing storm water drainage systems for new development. Detention facilities are frequently required in subdivision laws, zoning ordinances, building codes, and water pollution regulations.

—*Kent K. Mao*

6.2
NONPOINT SOURCE POLLUTION

Urban storm water pollution and most pollution in combined sewer overflows originates from nonpoint or diffuse sources. The processes controlling storm water quality are rather complex, as shown in Figure 6.2.1. In contrast to point source pollution, such as industrial and municipal treatment plant outfalls, these sources of pollution are numerous and their contributions are difficult to quantify. Diffuse pollution is a hydrologic process that closely follows the statistic character of rainfall, and must be evaluated similarly.

Urban nonpoint sources have been identified as a major cause of pollution of surface water bodies by the U.S. EPA (EPA 1984). In the 1988 Report to Congress (EPA 1990), the EPA stated that urban storm water runoff is the fourth most extensive cause of impaired water quality in the nation's rivers, and the third most extensive cause of impaired water quality in lakes. Combined sewer outflows (CSOs) are tenth on the list of significant sources of impairment for both surface-water bodies.

Major Types of Pollutants

Urban storm water runoff may transport many undesirable pollutants. The pollutants present, and their concentrations, are a function of the degree of urbanization, the type of land use, the densities of automobile traffic and animal population, and the degree of air pollution before rainfall. Major pollutant types are classified as follows:

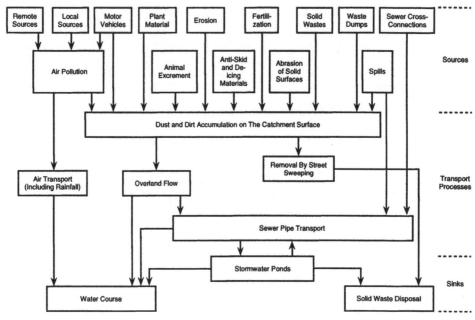

FIG. 6.2.1 Sources of pollutants in stormwater and pollutant pathways. (Reprinted from United Nations Educational, Scientific and Cultural Organization (UNESCO) 1987, Manual on drainage in urbanized areas, vol. 1, *Planning and design of drainage systems,* UNESCO Press.)

- Suspended sediments
- Oxygen-demanding substances
- Heavy metals
- Toxic organics—pesticides, PCBs
- Nutrients—nitrogen, phosphorous
- Bacteria and viruses
- Petroleum-based substances or hydrocarbons
- Acids and
- Humic substances—precursors for trihalomethane

Annual pollutant loadings for storm water and combined sewer overflows are given in Table 6.2.1 (UNESCO 1987).

Nonpoint Sources

Three basic processes generate pollutants during runoff:

ATMOSPHERIC DEPOSITION

Atmospheric deposition is generally divided into wet and dry deposition. Wet deposition is closely related to the levels of atmospheric pollution by traffic, industrial and domestic heating, and other sources. Urban rainfall is generally acidic, with below 5 pH units. The elevated acidity of urban precipitation damages pavements, sewers, and other

TABLE 6.2.1 ANNUAL UNIT POLLUTANT LOADINGS FOR STORMWATER AND COMBINED SEWER OVERFLOWS

	Annual Pollutant Loadings (kg/ha/yr)*				
Source	Total Suspended Solids	BOD	COD	Total N	Total P
Runoff in storm sewers	100–6300	5–170	20–1000	2–12	0,2–2,2
Residential area runoff	600–2300	5–100	20–800	2–12	0,2–2,2
Commercial area runoff	100–800	40–90	100–1000	5–12	1,2–2,2
Industrial area runoff	400–1700	10–90	200–1000	5–10	1,0–2,1
Highway runoff	120–6300	90–170	180–3900	—	—
Combined sewer overflows	1200–5000	500–1300	500–3300	15–40	4–8

*1 kg/ha/yr = 0.89 lb/acre/yr

Source: Reprinted from United Nations Educational, Scientific and Cultural Organization (UNESCO), 1987, Manual on drainage in urbanized areas, vol. 1, *Planning and design of drainage systems,* (UNESCO Press).

building materials. Particles are then washed off the surface by stormwater.

Dry atmospheric deposits are fine particles originating from a distance (fugitive dust) or locally (traffic on unpaved roads, construction and industrial) sources. Dustfall rates vary from region to region. Rural dustfalls depend on soil condition; urban dustfalls are more related to local air pollution.

EROSION

Erosion of construction areas represents the largest source of sediments in urban runoff. Reported unit loads of sediment from urban construction sites ranged from 12 to 500 tons/half-yr (Novotny and Chesters 1981). Furthermore, building activities generate other pollutants such as chemicals from fertilizers and pesticides, petroleum products, construction chemicals (cleaning solvents, paints, acids and salts), and various solids. Grading exposes subsoil, increasing surface erosion due to stormwater runoff.

Erosion of urban lawns and park surfaces is usually low. Exceptions are open, unused lands, and construction sites.

Soil is a source of suspended solids, organics, and pesticide pollution. Despite the SCS's active promotion of erosion control, the U.S. Department of Agricultural estimates 57–76 million acres (21–31 Mha), about 15–25% of the nation's agricultural land, is in need of sediment control measures.

ACCUMULATION/WASHOFF

Most urban watersheds are dominated by accumulation and wash-off processes, depending on impervious areas. The accumulation of solids on impervious urban surface areas is described by Sartor and Boyd (1972), as shown in Figure 6.2.2.

The sources of urban diffused pollution are:

- Litter, including large-sized materials (greater than 3.2 mm) containing items such as cans, broken glass, vegetable residues and pet waste. Pet fecal deposits can reach alarming proportions in urban centers where large numbers of people reside in highly impervious zones.
- Medium size deposits (street dirt) represent the bulk of street surface-accumulated pollution. The sources are numerous and very difficult to identify and control. They may include traffic, road deterioration, vegetation resides, pets and other animal waste and residues, and decomposed litter.
- Traffic emissions are responsible for potentially toxic pollutants found in urban runoff, including lead, chromium, asbestos, copper, hydrocarbons, phosphorous, and zinc. Pollution also comes from particles of rubber abraded from tires.
- Road deicing salts applied in winter cause highly increased concentrations of salts in urban runoff.

Road salts are applied at rates of 100–300 kg/km of highway and contain sodium and calcium chloride.
- Pesticides and fertilizers applied onto grassed urban lands.

In fully developed urban areas, where most land surfaces are impervious because of paving and rooftops, washoff of deposited particles and their transport to the watercourse become the important mechanism. The relationship of imperviousness to the quantity of some pollutants are shown in Table 6.2.2.

Table 6.2.3 shows values and ranges of accumulation of street and surface pollutants estimated by Ellis (1986). A list of specific nonpoint sources is presented in Table 6.2.4. The list is not exhaustive. The importance of the sources varies with local conditions.

Direct Input from Pollutant Source

Nonpoint pollutants can also reach receiving waters by direct input from a pollutant source. Drainage systems include depressions, ditches, culverts, catch basins, wetlands, and creeks that collect water and transport pollutants to receiving waters. Pollutants may be directly introduced at specific sites in the system, independent of storm conditions. For example, substances may be poured into a catch basin, traveling directly into a creek or other receiving water.

In addition to cross-connections of sewage and industrial wastes from sanitary sewers, solid waste dumps, and failing septic tanks, solids accumulations and growth in sewers can also enter into storm sewers. Excess water from lawn watering and car washing is another example of direct input. Pollutant loadings from direct inputs are difficult to document and quantify.

—Kent K. Mao

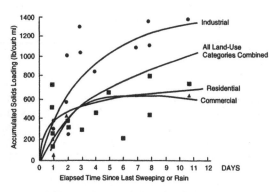

FIG. 6.2.2 Pollutant Accumulation for Different Urban Land Uses. (Reprinted, with permission from J.D. Sartor and S. Boyd, 1972, *Water pollution aspects of street surface contaminants,* U.S. Environmental Protection Agency (EPA), EPA Report R2–72–087, Washington, D.C.)

TABLE 6.2.2   ANNUAL STORM POLLUTANT EXPORT FOR VARIOUS LAND USE TYPES BY PERCENT IMPERVIOUS COVER (POUNDS/ACRE/YEAR)

General Land Use	Percent Imperviousness	Total Phosphorous	Total Nitrogen	BOD 5-Day	Zinc	Lead
Rural to	0	0.11	0.80	2.10	0.02	0.01
residential	5	0.20	1.60	4.00	0.03	0.01
	10	0.30	2.30	5.80	0.04	0.02
Large lot,	10	0.30	2.30	5.80	0.04	0.02
single family	15	0.39	3.00	7.70	0.06	0.03
	20	0.49	3.80	9.60	0.07	0.04
Medium density	20	0.49	3.80	9.60	0.07	0.04
single family	25	0.58	4.50	11.40	0.08	0.05
	30	0.68	5.20	13.30	0.10	0.05
	35	0.77	6.00	15.20	0.11	0.06
Townhouse	35	0.77	6.00	15.20	0.11	0.06
	40	0.86	6.70	17.10	0.12	0.07
	45	0.97	7.40	18.90	0.14	0.07
	50	1.06	8.20	20.80	0.15	0.08
Garden apartment	50	1.06	8.20	20.80	0.15	0.08
buildings	55	1.16	8.40	22.70	0.16	0.09
	60	1.25	9.60	24.60	0.18	0.09
High rise to light	60	1.25	9.60	24.60	0.18	0.09
commercial/industrial	65	1.35	10.40	26.40	0.19	0.10
	70	1.44	11.10	28.30	0.21	0.10
	75	1.54	11.80	30.20	0.22	0.11
	80	1.63	12.60	32.00	0.23	0.11
Heavy commercial to	80	1.63	12.60	32.00	0.23	0.11
shopping center	85	1.73	13.30	33.90	0.25	0.12
	90	1.82	14.00	35.80	0.26	0.13
	95	1.92	14.80	37.70	0.27	0.13
	100	2.00	15.40	39.20	0.28	0.14

NOTES: Assumed rainfall of 40 in/yr
Rural residential = 0.25–.5 dwelling units/acre
Large lot, single family = 1–1.5 dwelling units/acre
Medium density, single family = 2–10 dwelling units/acre
Townhouse and garden apartment = 10–20 dwelling units/acre
Pollutant loadings are for new developments only.

TABLE 6.2.3   SOLIDS ACCUMULATION AND ASSOCIATED POLLUTANT CONCENTRATIONS IN URBAN AREAS

Land Use		Residential Low Density	High Density	Commercial	Light Industrial	Highways
Solids accumulation (g/curb m)		10–182	30–210	13–180	80–288	13–1100
Pollutant concentration (μg/g)	BOD_5	5260	3370	7190	2920	2300–10,000
	COD	39,300–40,000	40,000–42,000	39,000–61,730	25,100	53,650–80,000
	Tot.N	460–480	530–610	410–420	430	223–1600
	Pb	1570	1980	2330	1390	450–2346
	Cd	3.2	2.7	2.9	3.6	2.1–10.2
Fecal Coliforms (MPN/g)		60,570–82,500	25,621–31,800	36,900	30,700	18,768–38,000

Source: Reprinted, with permission, from J.B. Ellis Pollutional aspects of urban runoff, *Urban runoff pollution,* ed. H.C. Torno, J. Marsalek, and M. Desbordes, 1–38. (New York, N.Y.: Springer, Verlag, Berlin).

TABLE 6.2.4 NONPOINT SOURCE POLLUTANTS

Source	N	O/G	T	S	O	M	P	H
Agricultural								
Nurseries	X		X	X	X			
Crop farms	X		X	X	X			
Livestock/hobby farms	X			X	X		X	X
Feed/seed/fertilizer supply	X		X		X			
Commercial/Retail								
Restaurants	X	X			X		X	
Dry cleaners			X					X
Garden centers	X		X	X	X	X		
Printing shops			X					
Urban Stormwater								
Roof washoff	X	X	X	X	X	X		X
Lawn/landscape washoff	X		X	X	X		X	
Yard debris	X			X	X		X	
Septic systems	X		X		X		X	
Household		X	X			X		
Miscellaneous								
Illicit dumps	X	X	X	X	X	X	X	
Cemeteries	X		X	X				
Warehouses		X	X		X	X		X
Fuel storage facilities		X	X			X		
Streambank erosion	X			X	X			
Ditch cleaning/defoliating	X		X	X	X			
Filling/diverting streams	X			X	X			
Loss of buffer zones				X	X			X
Boating and marinas	X	X	X		X		X	
Construction								
Clearing/grading	X	X		X	X			X
Building		X	X	X				
Transportation								
Roadways/parking lots	X	X	X	X	X	X		X
Service/repair stations		X	X	X	X	X		X
Car/truck washes	X	X	X	X	X	X		X
Oil change shops		X	X	X	X	X		X

N = nutrients; O/G = oils and greases; T = toxic chemicals; S = sediments; O = organics; M = metals; P = pathogens, bacteria; H = heat

References

Browne, F.X. and J.T. Grizzard. 1979. Nonpoint sources, J. Water Pol. Cont. Fed. 51, p. 1428.

Ellis, J.B. 1986. Pollutional aspects of urban runoff, in *Urban runoff pollution*, eds. H.C. Torno, J. Marsalek, and M. Desbordes, 1–38. New York, N.Y.: Springer Verlag, Berlin.

Novotny, V. and G. Chesters. 1981. *Handbook of nonpoint pollution: source and management.* New York, N.Y.: Van Nostrand Reinhold.

Rechow, K.H., M.N. Beaulac, and J.T. Simpon. 1980. *Modeling phosphorous loading and lake response under uncertainty: A manual and compilation of export coefficients.* U.S. Environmental Protection Agency (EPA) EPA 440–5–80–011. Washington, D.C.

Sartor, J.D. and G. Boyd. 1972. *Water pollution aspects of street surface contaminants.* U.S. Environmental Protection Agency (EPA) EPA R2–72–081. Washington, D.C.

United Nations Educational, Scientific and Cultural Organization (UNESCO). 1987. Manual on drainage in urbanized areas. vol. I. *Planning and design of drainage systems.* UNESCO Press.

U.S. Environmental Protection Agency (EPA). 1990. *National water quality inventory—1988 report to Congress.* U.S. EPA Office of Water. EPA Report 440–4–90–003. Washington, D.C.

6.3
BEST MANAGEMENT PRACTICES

Much emphasis is currently placed on controlling storm water pollution by attacking the problem at the source, instead of using more costly downstream treatment facilities. These source controls, termed "Best Management Practices" (BMPs), are judged most effective in reducing nonpoint source pollution to a level compatible with water quality goals.

Best Management Practices are classified into two groups:

- Planning, with efforts directed at future development and redevelopment of existing areas
- Maintenance and operational practices to reduce the impact of nonpoint source contamination from existing developed areas

Successful storm water pollution control depends on the effective implementation of proposed planning efforts and/or control practices. Legislation or ordinances encouraging or requiring conformance with intended BMPs has proven to be effective. Table 6.3.1 lists activities included in a typical source control program.

Planning

The first goal of planning is to develop a macroscopic management concept, preventing problems from short-sighted development of individual areas. The planner is interested in controlling storm water volume, rate, and pollutional characteristics of storm water runoff. Since the size of storm sewer networks and treatment plants relates directly to flow quantity, particularly the peak flowrate, reducing total volume or smoothing out peaks will result in lower construction costs.

TABLE 6.3.1   TYPICAL SOURCE CONTROLS

Activity or Area	Overview
S1.10—Fueling stations (both commercial and private)	Covered, concrete-paved pump island with separate drainage
S1.20—Vehicle/equipment wash and steam cleaning	Wash building or designated paved area with separate drainage containing oil-water separator
S1.30—Loading and unloading liquid materials	Conduct activities inside building or at dock with overhang or skirts to prevent drainage to storm drains; rail and tanker truck transfers require drip pans or paved areas, and operations and spill cleanup plans
S1.40—Above-ground tanks for liquid storage	Diked secondary containment area with stormwater drainage passing through oil-water separator
S1.50—Container storage of liquids, food wastes, or dangerous wastes	Containers kept indoors or under designated covered area with separate drainage and secondary containment for liquid wastes
S1.60—Outside storage of raw materials, by-products, or finished product (i.e., sand and gravel, lumber, concrete and metal)	Place materials under covered area, place temporary plastic sheeting over material, *or* pave the area and install treatment drainage system
S1.70—Outside manufacturing	Alter, enclose, cover, or segregate the activity; discharge runoff to sewer or process wastewater system; or use stormwater BMPs
S1.80—Emergency spill cleanup plans	Required for storing, processing, or refining oil products and producers of dangerous wastes
S1.90—Vegetation management /integrated pest management	Specific BMPs for seeding and planting, and pest management, including use of pesticides
S2.00—Maintenance of storm drainage facilities	Specific BMPs for maintenance (inspection, repair, and cleaning), disposal of contaminated water, and disposal of contaminated sediments

Source: Stormwater Management Manual for the Puget Sound Basin (Ecology 1992).

LAND USE PLANNING

Computer simulations are used to examine interacting pollutant sources in the watershed. By modeling the runoff process, a planner can predict the effects of proposed plans, and the ability of controls to solve potential problems. Water quality criteria standards can be recommended after investigating pollution sources and the ability of receiving water to absorb loadings.

When watershed goals are set, the planning agency has two choices for achieving water quality standards. Individual sites can be forced to comply with the practices and performance standards set forth in the master plan, or the basin system must be designed and maintained as a public utility. Isolated development tracts can be controlled by requiring developers to follow specific source control practices, or a simple set of performance standards can be applied and the choice of practices can be left up to the developer. For example, the agency can require that runoff from developed sites must not exceed predevelopment intensity. The developer will have to minimize runoff-producing areas and provide detention facilities at the site.

Planners must also consider the effects of their actions on areas outside the watershed. For example, a system where storm flow is detained in a downstream watershed while it remains unregulated upstream can cause higher flood levels in a river than a completely unregulated system.

NATURAL DRAINAGE FEATURES

The key to preserving a natural drainage system for urbanizing areas is understanding the predevelopment water balance and designing to minimize interference with that system. The soil and hydrology of the site must be studied so that high-density, highly impervious locations, such as shopping centers and industrial complexes, are located in areas with low infiltration potential. Recharge areas should be preserved as open, undisturbed space in parks and woodlands. Runoff from developed areas should be directed to recharge areas and detained to use the full infiltration potential. Broad, grassy swales will slow runoff and maximize infiltration. The drainage plan can include variable depth detention ponds that rise during a runoff event and return to a base level during dry weather.

Realizing that the design goal is maximizing infiltration-recharge and minimizing runoff, the planner should incorporate the following techniques into a site plan:

- Roof leaders should discharge to pervious areas or seepage pits. Dry (French) wells, consisting of borings filled with gravel, can be used for infiltration of rooftop runoff.
- As much area as possible should be left in a natural, undisturbed state. Earthwork and construction traffic will compact soil and decrease infiltration.
- Steep slopes should be avoided. They contribute to erosion and lessen recharge.
- Large impervious areas should be avoided. Parking lots can be built in small units and drained to pervious areas.
- No development should be permitted in flood plains.

Porous pavement is an alternative to conventional pavement. (Thelen and Howe 1978; Dinitz 1980). It provides storage, enhancing soil infiltration to reduce surface and volume from an otherwise impervious area.

For parking lots and access roads, planners can use *modular pavement* systems. Pavers are placed on a prepared sand and gravel base, which overlays the subsoil. The voids of the pavers are filled with either sand, gravel, or sod. Frost problems are minimal.

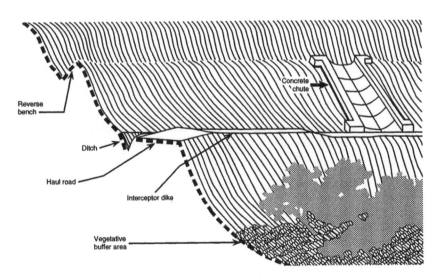

FIG. 6.3.1 Interception and diversion measures.

TABLE 6.3.2 SELECTING BMPs BY POLLUTANTS

Methods of Control	Structural	Vegetative	Management
Sediment (TSS, cobble embeddedness, turbidity)			
Control erosion on land and streambank	Terraces; diversions; grade stabilization structures; streambank protection and stabilization	Cover crops and rotations; conservation tillage; critical area planting	Contour farming; riparian area protection; proper grazing use and range management
Route runoff through BMPs that capture sediment	Sediment basins	Filter strip; grassed waterway; stripcropping; field borders	
Dispose of sediment properly			Beneficial use of sediment—wetland enhancement
Nutrients: N, P (nuisance algae, low dissolved oxygen, odor)			
Minimize sources	Animal waste system (lagoon, storage area); fences (livestock exclusion); diversions; terraces	Range management; crop rotations	Range and pasture management; proper stocking rate; waste composting; nutrient management
Uptake all that is applied to the land or contain and recycle/ reuse (dissolved form control—commercial nutrients)	Terrace; tailwater pit; runoff retention pond; wetland development	Cover crop; strip cropping; riparian buffer zone; change crop or grass species to one that is more nutrient demanding	Recycle/reuse irrigation return flow and runoff water; nutrient management; irrigation water management
Contain animal waste, process and land apply, or export to a different watershed (dissolved form control—animal waste)	Diversion; pit/pond/lagoon; compost facility	See 2(a)	Lagoon pump out; proper irrigation management
Minimize soil erosion and sediment delivery (adsorbed form control)	Terrace; diversion; stream-bank protection and stabilization; sediment pond; critical area treatment	Conservation tillage; filter strip; riparian buffer zone; cover crop	Nutrient management
Intercept, treat runoff before it reaches the water (suspended form control)	See 1–3; water treatment (filtration or flocculation) for high-value crops	Riparian buffer zone	*See* four preceding items, this column
Pathogens (bacteria, viruses, etc.)			
Minimize source	Fences		Animal waste management, especially proper application rate and timing
Minimize movement so bacteria dies	Animal waste storage; detention pond	Filter strips; riparian buffer zones	Proper site selection for animal feeding facility; proper application rate of waste
Treat water	Waste treatment lagoon; filtration	Artificial wetland/rock reed microbial filter	Recycle and reuse
Metals			
Control soil sources		Crop/plant selection	Avoid adding materials containing trace metals

Continued

TABLE 6.3.2 *Continued*

Methods of Control	Structural	Vegetative	Management
Control added sources	Tailwater pit; reuse/recycle system	Crop selection	Irrigation water management; integrated pest control
Treat water	Filtration	Artificial wetland/rock reed microbial filter system	
Salts/salinity			
Limit availability			Drip irrigation
Control loss	Evaporation basins; tailwater recovery pits; ditch lining; replace ditches with pipe	Crop selection; saline wetland buffer; land-use conversion	Irrigation water management
Pesticides and other toxins			
Minimize sources		Plant variety/crop selection	IPM; change planting dates; proper container disposal
Minimize movement and discharge	Terrace; sediment control basin; retention pond with water reuse/recycle system	Buffer zone; conservation tillage; filter strips (adsorbed control only); wetland enhancement	Irrigation water management; IPM
Treat discharge water	Carbon filter system (high-value crops)	Rock-reed microbial filter system/artificial wetland	
Physical habitat alteration			
Minimize disturbance within 100 feet of water	Road and turnrow realignment; fencing/livestock water crossing facility	Buffer strips; riparian buffer zones	Proper grazing management, including limiting livestock access
Control erosion on land	See sediment BMPs		
Maintain or restore natural riparian area vegetation and hydrology	Streambank stabilization; channel integrity repair	Wetland enhancement	Proper grazing use and range management; limit livestock access

Sources: U.S. EPA (1993); Brach (1990); Alexander (1993a); USDA, Soil Conservation Service (1988).

EROSION CONTROLS

Erosion control for construction and developing sites will have a major impact on the total pollution loads in receiving waters. Current estimates show that approximately 1500 sq mi of the United States is urbanized annually. All of this land area is exposed to accelerated erosion.

Following are basic guidelines and principles of erosion control. Reduce the area and duration of soil exposure. For example, various mining operation stages should be scheduled so that clearing, grubbing, scalping, grading and revegetation occur concurrently with extraction, so that a minimum area is exposed at one time.

Protect the soil with mulch and vegetable cover. For example, covering the soil surface with wood chips reduces construction site soil loss by 92%. Vegetation also has a marked effect on water quality. Temporary fast-growing grass can reduce erosion by an order of magnitude; sod-

ding can reduce erosion by two orders of magnitude. Straw mulch application can be combined with grass seeding for permanent surface protection.

Reduce the rate and volume of runoff by increasing the infiltration rate. A properly roughened and loosened soil surface will benefit plant growth, enhance water infiltration, and slow surface runoff.

Diminish runoff velocity with planned engineering works. A key concept in controlling soil erosion is to intercept runoff before it reaches a critical area and divert it to a safe disposal area. Interception and diversion are accomplished through various structures, including earth dikes, ditches, and combined ditch and dike structures (Figure 6.3.1).

Protect and modify drainage ways to withstand concentrated runoff from paved areas. To reduce the rate of flow and the resulting detachment and transport of soil particles in natural and manmade drainageways, grade can be controlled by the construction of flumes or other flow barriers across the channel. Bends in the channel, either natural or manmade, also impede flow.

Trap as much sediment as possible in temporary or permanent sedimentation basins.

Maintain completed works and assure frequent inspection for maintenance needs.

Principal cropland erosion control practices and BMPs for pollutants are summarized in Table 6.3.2.

Maintenance and Operational Practices

Proper maintenance and cleanliness of an urban area can have a significant impact on the quantity of pollutants washed from an area by storm water. Cleanliness of an urban area starts with control of litter, debris, deicing agents, and agricultural chemicals such as pesticides and fertilizers. Regular street repair and sweeping can further minimize pollutants in stormwater runoff. Proper drainage collection system use and maintenance can maximize control of pollutants by directing them to treatment or disposal.

URBAN POLLUTANT CONTROL

Litter Control

Used food containers, cigarettes, newspapers, sidewalk sweepings, lawn trimmings, and other materials carelessly discarded become street litter. Unless this material is prevented from reaching the street or is removed by street cleaning, it is often found in stormwater discharges. Enforcement of antilitter laws, convenient location of disposal containers, and public education programs are source control measures.

Chemical Use Control

Reducing the indiscriminate use and disposal of fertilizers, pesticides, oil and gasoline, and detergents is a frequently overlooked measure for reducing stormwater runoff pollution. Tree spraying, weed control, municipal fertilization of parks and parkways, and homeowner use of pesticides and fertilizers can be controlled by increasing public awareness of the potential hazards to receiving waters. Direct dumping of chemicals and debris into catch basins, inlets, and sewers is a significant problem that can only be addressed through educational programs, ordinances, and enforcement.

Street Sweeping

Street sweeping is used by most cities to remove accumulated dust, dirt and litter from street surfaces, but clean-ing is usually done for aesthetic reasons. Street cleaning practices effectively attack the source of stormwater-related problems.

The type of cleaning equipment has an effect on the overall effectiveness of debris removal. Public awareness of street cleaning practice is essential for more efficient operations. Vehicles parked on the street during sweeping operations hamper efficiency and prevent cleaning of deposits.

Street Maintenance

Pavement conditions have an effect on the amount of street pollutants. Vehicles travelling over rough streets shake off more particulate matter. A large portion of solids also comes from cracks in the pavement.

Highway Deicing Management

Effective management of highway deicing practices can lessen receiving water impacts associated with chlorides, sodium, and suspended solids. Recommended alternatives for modifying deicing practices include: (1) judicious application of salt and abrasives; (2) reducing application rates using sodium and calcium salt premixers; (3) using better spreading and metering, and calibrating application rates; (4) prohibiting use of chemical additives; (5) providing improved salt storage areas; and (6) educating the public and operators about the effect of deicing technology and best management practices.

COLLECTION SYSTEM MAINTENANCE

The major objective of maintaining storm or combined sewer systems is to provide maximum transmission of flows to treatment and disposal, while minimizing overflows, bypasses, and local flooding conditions. This objective can be achieved by maintaining system facilities at peak capacity.

The significance of collection system maintenance as a best management practice is that when properly applied, extraneous solids and debris are removed in a controlled manner, not accumulated as pollutant sources to be flushed into receiving waters under storm conditions.

The basic part of a maintenance program is regular system inspection. Specific tasks include: (1) catchbasin maintenance; (2) cleaning (both deposits and root infestation) and flushing of pipes; (3) removal of excess shrubbery and debris from flood control channels and ditches; and (4) control of inflow and infiltration sources.

Sewer cleaning involves routine inspection of the sewer system. All plugged or restricted lines should be cleaned. Major problems in large-diameter sewers are siltation and accumulation of large debris like shopping carts and tree branches. In small-diameter sewers, siltation and penetration of tree roots are major problems. Benefits of sewer cleaning include reducing local flooding, emergency repairs, and pollutant loading. Increased carrying capacities and reduced blockages in interceptor/regulator works may directly reduce overflows.

Many types of sewer cleaning equipment are used, including hydraulic, mechanical, manual, and combination devices. The cleaning tool is pushed or pulled through the sewer to remove obstructions or cause them to be suspended in the flow and carried out of the system. However, large sewer and interceptor cleaning involves unique problems because several feet of sludge blanket can accumulate.

Regular flushing of sewers can ensure that sewer laterals and interceptors continue to carry their design capacity, as well alleviate solids buildup and reduce solid overflow.

Sewer flushing can be particularly beneficial in sewers with very flat slopes. If a modestly large quantity of water is periodically discharged through these flat sewers, small accumulations of solids can be washed from the system. This cleaning technique is effective only on freshly deposited solids.

Internally automatic flushing devices have been developed for sewer systems. An inflatable bag is used to stop flow in upstream reaches until a volume capable of generating a flush wave is accumulated. When the correct volume is reached, the bag is deflated with the assistance of a vacuum, releasing impounded water and cleaning the sewer segment.

INFLOW AND INFILTRATION

Extraneous flow entering a sewer is generally categorized as inflow or infiltration. Inflow generally occurs from surface runoff via roof leaders, yard and area away drains, and flooding of manhole covers. Infiltration usually occurs by water seeping into pipes or manholes from leaky joints, crushed or collapsed pipe segments, leaky lateral connections or other pipe failures. Extraneous flows may result in unnecessary pollution, as these reduce effective collection system and treatment plant facilities.

Details of principal methods of reducing both infiltration and inflow through rehabilitation are found in EPA 1977.

DRAINAGE CHANNEL MAINTENANCE

Maintenance of flood control channels covers a wide range of cleaning tasks. Debris to be removed ranges from trash, garbage, and yard trimmings to used tires and shopping carts.

—Kent K. Mao

References

Dinitz, E.V. 1980. *Porous pavement. Phase I. Design and operational criteria.* U.S. Environmental Protection Agency (EPA). EPA Report 600–2–80–135.

Stewart, B.A., D.A. Woodhiser, W.H. Wischmeier, J.H. Caro, and M.H. Frere. 1975. *Control of water pollution from cropland. Vol. I.* U.S. Environmental Protection Agency (EPA). EPA Report 600–2–75–026a. Washington, D.C.

Thelen, E. and L.F. Howe. 1978. *Porous pavement.* Philadelphia, Pa.: The Franklin Institute.

U.S. Environmental Protection Agency (EPA). 1977. *Sewer system evaluation, rehabilitation, and new construction: A manual of practice.* EPA Report 600–2–77–017d.

6.4
FIELD MONITORING PROGRAMS

The objectives of field monitoring water quality in drainage studies include:

- Analyzing the impact on receiving waters of (1) storm sewer discharges, (2) combined sewer overflows, (3) atmospheric fallout and urban activities, and (4) new facilities or treatment plants designed to reduce environment impacts.
- Identifying the contributions of various land uses to total pollution discharge, to optimize urban development and derive some regulations such as source control.

- Increasing existing treatment efficiency during wet weather in combined sewer systems.
- Analyzing of scour and deposit problems in sewers to define optimal cleaning sequences or to design facilities for better hydraulic conditions.

To fulfill these objectives, storm water discharges need to be sampled during dry-weather and wet-weather conditions. Water quality data gathered during dry weather provide a baseline and indicate point source impacts.

To trace contaminants and identify pollutant sources, a phased monitoring approach requires repeated investigation of land use activities in a basin. The program is expected to be an iterative process, as several rounds of sampling are generally required. Precise data are essential for calibrating and verifying nonpoint source models.

Experience proves that water quality data collection programs can be costly. Collection procedures have high manpower requirements, as frequent site visits are required. The cost of analyzing collected samples may increase rapidly with the number and types of pollutants studied. It is important that the parameters to be studied are carefully selected and limited to the essentials.

This section presents an outline of water quality parameters important in studies on urban stormwater discharges, and reveals the main difficulties in obtaining representative samples. Also included is a brief discussion on data analysis.

Selection of Water Quality Parameters

Water quality parameters included in urban hydrological studies may be divided into seven groups. Those parameters, relating to a specific drainage problem, are listed in Table 6.4.1, along with their detection limits, precision level of analysis, and study objectives. In most cases, only biochemical oxygen demand (BOD), chemical oxygen demand (COD), and total suspended solids (TSS) are initially studied, but if these parameters show high values, some other parameters can be taken into account (i.e. Kjeldahl nitrogen, total phosphorous, and volatile suspended solids [VSS]). As the program continues, some special investigation should be made on trace elements and other special parameters mentioned in Table 6.4.1.

Solids are good indicators of urban water quality, as they may contain pollutant materials. Suspended solids are closely related to other pollutant concentrations. In fact, sample uniformity is not easily achieved. Suspended solid concentrations are affected by the flow level, which is not taken into account by manual or automatic sampling. The sampler itself may also introduce effects that can modify the gradient profile of suspended solids. Conditions at the sampler intake cannot be adapted to the flow variations encountered in storm sewers or combined sewers during high flows.

If the sampler cannot be precisely measured in the collected samples, sample uniformity can be questionable. In most cases, suspended solids are regarded as a rough indicator of water quality, so this should be among the parameters selected.

Acquisition of Representative Samples

The number of sampling sites, the frequency of measurements, and the quality parameters to be measured should

be chosen. This requires knowledge of the sewer network, significant building activities, street cleaning practices, atmospheric pollution sources affecting the experimental sites, erosion patterns in surrounding natural areas, industrial activities, seasonal or climatic changes, etc., in order to avoid erroneous judgements in understanding the phenomena studied.

The experimental design must be in agreement with the study objectives (Geiger 1981, Gideometeozdat 1984a, Wong and Marsalek 1981). However, trial and error procedures should be used at the beginning of the study for a few basic parameters (for example, BOD, COD, TSS) at a few sampling sites. This information should be used when determining the experimental design.

SAMPLING SITES AND LOCATION

Sampling sites must be chosen according to study objectives, but hydraulic conditions and constraints necessary to the adopted procedures should be given attention. The sampling site must be located at a section downstream of the study site, i.e. corresponding to well-known sewer systems, land use types, special activities, etc. It is recommended that highly turbulent sections with well mixed flow be sampled. However, for the study of sediment transport deposit, these conditions may not be suitable, as suspended solids in the highly turbulent section may be scattered.

For monitoring in-stream impacts, the area of interest should be bracketed by upstream and downstream stations. A control station on a hydrologically similar but undisturbed watershed can be used to determine baseline conditions.

Two types of monitoring stations are employed for nonpoint source surveys:

1. Small catchment stations ranging from 12 to 125 acres (5 to 50 ha) in size, are used to gather data on specific land uses or special areas. They are usually found on storm sewers, drainage ditches or small tributaries.

2. Another type of station is built to monitor larger basins of greater than 125 acres (50 ha), and measure nonpoint source pollution loads impacting a receiving body, such as stream channels or rivers.

There are cases where the final choice must be made from a group of catchments. In such cases, the technique of weighted suitability ratings, as developed for land use, is recommended (Alley 1977). Assignment of suitability values is perhaps the most subjective part of the schedule.

SAMPLING METHODS

Due to the transient nature of storm runoff phenomena, random collection of grab samples does not allow a true representation of pollutant transport. Even if grab samples are modified to concentrate on storm events, the error po-

TABLE 6.4.1   WATER QUALITY PARAMETERS SAMPLED IN URBAN DRAINAGE STUDIES

Parameters	Detection Limits	Precision Level (absolute or relative)	Study Objectives or Observations
Common Constituents and Indicators			
Chlorides		2.5–5%*	Impact of salts used for deicing
Water temperature		0.1°C–0.5°C	Cross-connections; parasites in waters
Conductivity (at 20°C)	$5\,\mu S/cm$	5%	Changes during runoff, monitoring and control
pH		0.05–0.1 unit*	Rainfall quality analysis
Turbidity			Sediment transport
Nutrients			
Kjeldahl nitrogen		0.1 mg/l	Impact on receiving waters
Total phosphorus		5–15%*	Eutrophication process
Ammonia	0.001 mg/l	1–10%*	Impacts on detention basins with recreational purposes
Nitrites and nitrates	0.05 mg/l	4.5–18%	Cross-connections
Organic Indicators			
BOD_5 (5 day BOD)		2–25 mg/l	Impact on receiving waters by oxygen depletion
COD (with <1.5 g/l chlorides)		1–5% (if COD >50 mg/l)	Cross-connections
Trace Elements			
Lead			Impact on receiving waters; toxics accumulation in sediments
Zinc and other heavy metals	$0.05–3\,\mu g/l$	2–10%†	
Solids			
TSS (at 105°C)	0.5 mg/l	2–5%	Turbidity, oxygen reduction, transport of toxics; increase of hydraulic roughness
VSS (at 550°C)	1 mg/l		Organic part, oxygen depletion
Settleable solids			Maintenance problems in sewers and detention basins in recreational areas
Bacterial Indicators			
Total coliforms			Impact on receiving waters with recreational use
Fecal coliforms			Detection of cross-connections
Special Parameters			
Persistent toxic substances (PTS) such as organochloride pesticides	$0.00005\,\mu g/l$	0.005–0.05 g/l†	Impact on receiving waters Pollution of receiving waters sediments
Polyaromatic hydrocarbons	$1–5\,\mu g/l*$		Bioaccumulation in food chains
Chlorinated benzenes	$0.002–0.02\,\mu g/l†$		

*Depending on the instrument and/or analysis method.
†Depending on the substance analyzed.

tential remains quite high because of variations in pollutant concentrations during runoff events.

Two basic methods can provide estimates of pollutant loading during a storm event:

1. To determine total pollution loading during a storm event, a *flow-weighted composite method* is adequate. In these methods, either aliquot volume or time between aliquots is varied to construct a truly flow-weighted composite from many samples. Analyzing the composite sample and using synoptic flow data allow computation of an accurate estimate of runoff pollution loads, if the intervals between samples are short.

2. When, in addition to total pollution loading, it is necessary to investigate load variations during a storm event, the *sequential discrete procedure* must be used. A

series of samples is retrieved during a monitored runoff event. Following laboratory analysis of each sample and analysis of synoptic flow data, the runoff hydrograph and a curve of pollutant concentration or loading as a function of time may be plotted as shown in Figure 6.4.1. By determining the area under the curve, an accurate estimate of the total pollutant load for an event may be determined.

The interval between sample collection for the above procedures depends on the response time and duration of the storm. In general, at least four samples on the rising limbs and six samples on the recessing limbs should be collected for proper resolution of nonpoint source pollution loads in urban areas.

Samples may be collected either manually or by automatic samplers. Table 6.4.2 shows a matrix of advantages and disadvantages related to each sampling technique. A summary of methods used in urban stormwater sampling and comments on each was prepared by Shelley (Shelley and Kirkpatrick 1975).

Experimental results show sediment distribution in a stream cross section flowing at 5 ft/sec. An analysis of water quality constituents in the stream cross-section should be made to determine the distribution across the width and from top to the bottom of the stream. Samples should be tested for a suspended parameter (such as TSS) and a soluble parameter (such as orthophosphate). The testing should be carried out at a small runoff event and a moderate-to-high flow event if possible. Vertical sampling should be done using depth samplers (such as Kenmeyer bottles) or closeable bottles if the stream is more than 4 to 5 ft (1.2 to 1.5 m) deep. This factor should be considered in designing manual and automatic sampling procedures.

FLOW MEASUREMENT

Flow measurement is perhaps one of the most important aspects of designing an urban collections program. No data collecting task will be capable of achieving its goals if the precision and accuracy of the flow data required for load calculations are not considered.

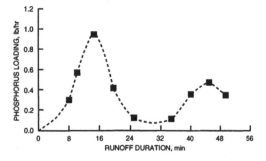

FIG. 6.4.1 Plot of total phosphorus loading at irongate catchment.

The flow measurement devices and methods can be classified according to the physical principles upon which their primary elements are based.

Channel Friction Coefficient Method

This indirect method, also referred to as the slope-area method, consists of measuring flow depth at a suitable cross-section and substituting the measured depth into an equation for uniform flow (such as the Manning equation) or critical flow. To complete the calculation, one must estimate the friction coefficient of the channel where the flow is to be measured, and know the channel slope and geometry.

The inference of flow rates from measured depths of flow is a rather inaccurate procedure. The main sources of error arise from the lack of uniformity and steadiness of flow, and the lack of certainty in estimating the friction coefficient.

Improved accuracy can be achieved by performing calibration in place, and developing an empirical rating curve for each measuring cross-section. In this case, the channel discharge (Q) is measured, generally by current meters, for various depths of flow, and the cross-section rating curve (Q vs depth of flow) is developed. This curve is then used to convert the observed stage to discharge.

Weirs

Measuring weirs are overflow structures built across a flow channel to measure discharge. For a given set of weir and channel geometry conditions, a single head value on the device may exist for each discharge under a free-flow, steady state regimen. The existence of such a relationship makes constructing a rating curve of head versus discharge a simple task. Such rating curves are available in the literature for most common configurations (such as rectangular weirs, V-notch weirs, vertical slot weirs, and trapezoidal weirs without the bottom part) (U.S. Department of Interior 1975).

One advantage of weirs is their large relative measurement range. However, weir installation in sewers reduces pipe capacity, may lead to solids accumulation (particularly in combined sewers), may distort flow hydrographs, and may limit operating range because of surcharging or submerging. These constraints will eliminate weirs from consideration for certain locations, but many of the above difficulties can be avoided in open-channel installations at outfalls. For these reasons, weirs should be used only under carefully controlled conditions, such as at detention basin outlets, where suspended solid concentrations are likely to be low.

Flumes

A measuring flume creates a constriction in the channel cross-section, causing a velocity change and, consequently,

TABLE 6.4.2 COMPARISON OF MANUAL AND AUTOMATIC SAMPLING TECHNIQUES

Advantages	Disadvantages
Manual Grabs	
Appropriate for all pollutants	Labor-intensive
Minimum equipment required	Environment possibly dangerous to field personnel
	May be difficult to get personnel and equipment to the storm water outfall within the 30 min requirement
	Possible human error
Manual Flow-Weighted Composites (multiple grabs)	
Appropriate for all pollutants	Labor-intensive
Minimum equipment required	Environment possibly dangerous to field personnel
	Human error may have significant impact on sample representativeness
	Requires flow measurements taken during sampling
Automatic Grabs	
Minimizes labor requirements	Samples collected for O&G may not be representative
Low risk of human error	Automatic samplers cannot properly collect samples for VOC analysis
Reduced personnel exposure to unsafe conditions	Costly, numerous sampling sites require the purchase of equipment
Sampling may be triggered remotely or initiated according to present conditions	Requires equipment installation and maintenance
	Requires operator training
	May not be appropriate for pH and temperature
	May not be appropriate for parameters with short holding times (e.g., fecal streptococcus, fecal coliform, chlorine)
	Cross-contamination of aliquot if tubing/bottles not washed
Automatic Flow-Weighted Composites	
Minimizes labor requirements	Not acceptable for VOC sampling
Low risk of human error	Costly if numerous sampling sites require the purchase of equipment
Reduced personnel exposure to unsafe conditions	Requires equipment installation and maintenance, may malfunction
May eliminate the need for manual compositing of aliquots	Requires initial operator training
Sampling may be triggered remotely or initiated according to on-site conditions	Requires accurate flow measurement equipment tied to sampler
	Cross-contamination of aliquot if tubing/bottles not washed

a depth change. In critical flow flumes, the surface profile in the constriction passes through the critical depth. The flume discharge can then be directly related to the depth immediately upstream of the throat.

Flumes are sometimes classified according to throat shape. Common types include rectangular, trapezoidal, semicircular, and composite throat flumes. Flumes with a bottom contraction (a hump) are suitable for installation in sewers. The Parshall flume, the cut-throat flume, and the Palmer-Bowlus flume are also popular.

Rating curves for critical flume geometry may be constructed from solution of the Bernouilli Equation at points upstream of and in the flume throat. While they generally exhibit excellent characteristics of self-cleaning, flumes do not share the brand flow measurement characteristics of weirs.

There are a large number of other flume designs that can be used in drainage studies. For example, the Soil Conservation Service has HS, H, and HL flumes designed to measure small, moderate, and large runoff flows, respectively. These devices combine the best features of both flumes and weirs, with wide ranges of measurement, good self-cleaning characteristics, small head loss, and relative insensitivity to submergence.

Differential Pressure Methods

Traditionally, differential pressure flowmeters have been used to measure flows in full closed conduit. Two exceptions to this rule are the U.S. Geological Survey and University of Illinois sewer meters. Although these function as differential meters in the pressure flow region, they are also fully functional in the open-channel flow region, where they act as Venturi flumes. This dual mode of operation represents the main advantage of these flowmeters.

The U.S. Geological Survey (USGS) flowmeter is similar to flumes with a U-shaped throat. The flume does not obstruct the part of the pipe immediately below the crown, thus transition from open-channel flow to pressure flow is fairly smooth and head losses are reduced. Rating curves for the USGS flowmeter are available (Smoot 1975).

Dilution Method

In this method, a tracer is continuously injected at a constant rate into the flow, and tracer dilution by the metered flow is monitored at a downstream point. If a tracer absent in the meter flow is used, the following relationship applies

$$Q_D = q_T C_T / C_D \qquad 6.4(1)$$

where:

Q_D = flow upstream
q_T = tracer input flow
C_T = tracer input concentration
C_D = tracer concentration downstream

The dilution method has some definite advantages, because it is independent of flow characteristics, does not interfere with the flow and, consequently, does not cause any head loss. Using fluorescent dyes and ensuring complete tracer mixing, the method has a good range of measurement (1000 : 1), and can be fairly accurate (5%) (Alley 1977). Disadvantages are the discrete nature of measurement, as opposed to the preferred continuous measurements; the problem with automating the method; and the need for well-trained personnel. Consequently, the dilu-

tion method is mostly used for in-situ calibration of conventional flowmeters.

Basic characteristics of flow measurement methods discussed are summarized in Table 6.4.3.

Sampling Equipment
MANUAL SAMPLING

Certain manual techniques cannot be avoided in studies of urban runoff quality. Manual sampling is useful when setting up automatic equipment, selecting the sampling section, and the inlet location.

Manual sampling requires good logistic preparation. Field crews must be dispatched to sampling sites before the start of a runoff event, so that sampling can start at the beginning of runoff. This is particularly important in combined sewers which exhibit the first flush phenomenon with high pollutant loads occurring early during runoff events. Therefore, field crews may have to be stationed at sampling sites. Extensive field training is essential to ensure collection of adequate samples.

AUTOMATIC SAMPLING

To obtain necessary flow measurements along with storm water samples, two devices are required: one for flow metering and one for flow metering with an interconnection to insure synoptic collection of sample and flow data. Common characteristics of adequate devices are summarized below:

- Sample transport velocity of 3.0 fps or more to prevent sedimentation
- Minimum of 24 discrete sample bottles or ability to composite samples in one container
- 12 v dc supply option
- Constant sample size over different sampling lines for rising and falling streams
- Air purging of sampling intake line before and after sample collection
- Minimum $\frac{3}{8}$ in (or 1 cm) sample line
- No solids deposition in sample train
- Chemically inert surfaces in contact with sample

In general, the intake should point upstream, extended slightly upstream from any obstacles in the flow, and should not excessively obstruct flow to avoid clogging or damage. Locations are recommended along the pipe periphery at about one third of the average water depth above the bottom. The intake should be placed at a cross-section where the flow is highly turbulent and well mixed. At such locations, a single intake, instead of multiple intakes, may be acceptable.

Sample withdrawal is accomplished by a pump controlled by timers or flow meters. The best devices for urban pollution studies fall into the following categories of pumping methods: positive displacement, peristaltic, and

TABLE 6.4.3   CHARACTERISTICS OF SELECTED FLOW MEASUREMENT METHODS

| | Characteristics | | | | | | |
| | Suitable For | | Applicable At | | | | |
	Open Channel Flow	Pressure Flow	Outfall	Manhole	In Sewer Pipes	Estimated Accuracy* (%)	Relative Costs
Depth and channel friction coefficient	X	X	X	X		15–20	low
Depth and stage-discharge relationship	X		X	X		10–15	low
Weirs							
Rectangular	X		X			5†	low to
V-notch	X		X			5†	medium
Modified trapezoidal	X		X	X	X	5†	
Vertical slot	X		X	X	X	5†	
Flumes							
Cut-throat	X		X			5†	
Palmer-Bowlus	X		X	X	X	5†	medium
Parshall	X		X			5†	
USDA (H, HL and HS)	X		X			5†	
Differential pressure flowmeters							
U. of Illinois	X	X			X	5	medium
USGS	X	X			X	5	to high
Tracer dilution	X	X		X	X	5	medium

*Under favorable conditions.
†These relatively high accuracies correspond to well-designed, installed, and operated installations. Under less favorable circumstances, the accuracies would be somewhat lower, between 5 and 10%.

vacuum. Suction lift devices are the best means of sample withdrawal. Such devices have to operate near the flow sampled because the lift is limited to about 15 ft (5 m). Submersible positive displacement pumps are commonly used where equipment installation is restricted to locations too high above the water surface to operate in a suction lift mode. Although such pumps allow sampling at greater depth, they are susceptible to malfunction and clogging.

FLOWMETERING DEVICES

Selection of secondary devices for the continuous measurements necessary to convert from stage to discharge is an important facet of developing an automated monitoring program. Important criteria for these secondary devices include:

- Wide measurement range
- Accuracy and precision over the entire range
- Minimal calibration loss with time
- Insensitivity to suspended solids in flow
- Capacity to internally convert stage to discharge
- Capacity to trigger an associated sampler
- Unattended operations

Secondary devices are divided into four categories: float-operated devices; ultrasonic devices; bubbler devices (manometers and transducers); and combination bubbler-magnetic devices.

Bubbler-Operated Devices

In the simplest of designs, a float is connected to a strip chart or digital recorder via flexible steel tape. In most applications, float-type devices require a stilling well to damp out surges and rapid fluctuations in water surface elevation. In addition, most float-operated devices do not provide an internal stage-to-discharge conversion.

Ultrasonic Devices

These secondary devices rely upon the travel time of an ultrasonic signal from a transponder to the water surface and back. This type of meter functions in a noncontact mode, and is therefore free from clogging and freezing. However, ultrasonics are sometimes subject to spurious signals from floating matter and foam. Some devices have internally programmable read-only memories (PROMs) and microprocessor circuitry to provide stage-to-discharge conversion using the unique relationships of the primary device.

Bubbler Devices

In bubbler devices, gas is forced through a fixed orifice, oriented to assure that only static head is measured. The static pressure required to maintain a given bubble rate is proportional to the height of the water column above the

TABLE 6.4.4   SUMMARY OF DATA ANALYSIS METHODS

Level of Analysis and Methods	Examples	References
Design of Experiments		

Factor analysis	Choosing experimental catchments or measuring sites for a given experimental program:
	    land uses
	    catchments parameters
	    water quality sampling
	Choosing number of experiments using physical models
		Cochran & Cox, 1957; Kendall & Stuart, 1973; Snedecor & Cochran, 1957

Raw Data Criticism		
Double mass analysis	Testing for systematic errors in time data series such as cumulative rainfall or runoff depths at various points in the same climatic areas	
Parametric tests (Anderson test)	Testing the random aspect of a data series such as rainfall and runoff	Bennet & Franklin, 1967; Dagnelie, 1970; Haan, 1977; Pearson & Hartley, 1969
Nonparametric tests Variance ratio test, Bartlett's test, et al.	Testing of the hypothesis on equal variance of two populations: rainfall runoff, runoff quality data	Dagnelie, 1970; Kendall & Stuart, 1973; Kite, 1976; Pettitt, 1979
Wilcoxon, Mann-Whitney, Kruskal-Wallis, Wilks tests	Testing of the hypothesis on equal means and identical location of population: rainfall or runoff, runoff quality data from several catchments	

Statistical parameters
  Arithmetic mean (or geometric mean for data lognormally distributed)
  Variance or standard deviation
  Ranges
  Pearson's and Fisher's coefficients
	Comparison of samples and homogeneity testing
	Parameters can be time and/or flow weighted for runoff quality data
	
	Preliminary statistical analysis
		All books on statistical methods

Point-Frequency Analysis		

Empirical frequency plotting
 	
  Probability papers
  Plotting formulae
	Analysis of a separate variable considered as a random variable:
	    rainfall depths for various time intervals (I.D.F. curves)
	    peak runoff
	    risk analysis
	Choice of a theoretical probability distribution
		Adamowski, 1981; Bennet & Franklin, 1967; Cunnane, 1973; Dagnelie, 1970; Haan, 1977; Kendall & Stuart, 1977a; Kite, 1976; Snedecor & Cochran, 1957; Yevjevich, 1972b

Theoretical probability (distributions discrete and continuous)
 	
  Method of moments
  Method of maximum likelihood
	Almost all hydrological variables (rainfall, runoff, quantity, quality) considered as a random variable	Chow, 1964; Dagnelie, 1970; Gumbel, 1960; Haan, 1977; Kendall & Stuart, 1977a; Kite, 1976; Linsley et al., 1975; Snedecor & Cochran, 1957; Viessman et al., 1977; Yevjevich, 1972b

Hypothesis testing and confidence intervals
 	
  Tests of means and variances
  Goodness-of-fit tests
	Testing the adequacy of a given probability distribution to a given sample	Chow, 1964; Dagnelie, 1970; Haan, 1977; Kendall & Stuart, 1977a; Kendall & Stuart, 1973; Kite, 1976; Snedecor & Cochran, 1957; Yevjevich, 1972b

Multivariate Analysis		

Simple Regression Analysis
  best fit procedure choice
  tests of fit
  spurious correlations
	Applied to a pair of hydrological variables
	    rainfall and runoff volumes
	    runoff coefficients and imperviousness
	    rainfall depths at two sites
	    overland flow detention storage and discharge
	    pollutants loads and peak runoff etc.
		Chatfield & Collins, 1980; Haan, 1977; Morrison, 1976; Draper & Smith, 1966; Haan, 1977; Viessman et al., 1977

Continued

TABLE 6.4.4 *Continued*

Level of Analysis and Methods	Examples	References

Multivariate probability distributions	Applied to several independent variables considered to be purely random variables
	    risk analysis in urban water management
	    spatial rainfall depths distribution
	    hydrological stochastic processes (discrete and continuous)
		Adamowski, 1981; Dagnelie, 1970; Kendall & Stuart, 1977a; Kite, 1976; Yevjevich, 1972a; Yevjevich, 1972b

Multiple regressions analysis
 	
Simple matrix procedure of best fit
Stepwise regression procedure
Orthogonal regression procedure
Ridge regression procedure
Cross validation procedure
Better results when Xi variables are correlated
	Applied to one explained variable Y and to several explanatory variables Xi:
	    interpolation between a set of raingauges
	    generation of data for incomplete data series
	    rainfall-runoff modeling at a given location versus rainfall and/or runoff at other locations
	    runoff coefficients versus urban catchment parameters and rainfall characteristics
	    lag times and times of concentration versus catchments and rainfall parameters
	    Urban runoff pollutant loads versus rainfall and runoff parameters, catchment characteristics such as land uses, imperviousness, slopes, etc.
		Chatfield & Collins, 1980; Draper & Smith, 1966; Haan, 1977; Pearson & Hartley, 1969; Robitaille and Bobbée, 1975; Stone, 1974; Yevjevich, 1972a

Interdependence analysis
 	
Correlation analysis
Principal components analysis (P.C.A.)
 	
Factor analysis
 	
Cluster analysis
Discriminant analysis
	Mostly for qualitative analysis. Not frequently applied in urban hydrology
	Just two variables
	More than two variables: reduction of dimensionality, preliminary analysis for regression procedures. Sometimes quantitative
	    spatial distribution of rainfall
	    urban runoff pollutants loads
	Similar aims as P.C.A. but with assumption of a proper statistical model. Covariance analysis
	Grouping tests of individuals
	Separation of individuals in two populations. Preliminary analysis for regression procedures
		Chatfield & Collins, 1980; Haan, 1977; Morrison, 1976

Time Series Analysis	Testing the random aspect of a given variable for preliminary statistical analysis Stochastic modelling of hydrological processes (not very frequent in urban hydrology due to time and space intervals to be considered)	Bartlett, 1966; Box & Jenkins, 1970; Cox & Miller, 1968; Jenkins & Watts, 1968; Kendall & Stuart, 1977b; Yevjevich, 1972a

Trend analysis
  Tests of randomness
  Least squares procedures
  Moving average methods
Periodic analysis
	Testing gradual natural or man-induced changes in data series
	    Changes in urban hydrological data due to continuous urbanization
	Testing the existence of cycles:
	    Seasonal aspects of rainfall, runoff, quality, quantity data
	    Short cycles due to some industrial or domestic water uses	

Spectral analysis on the time domain (Autocorrelation function) Spectral analysis on the frequency domain (Spectral density function)	Testing the random aspect of a given process Identifying Instantaneous Unit Hydrographs (IHU) for small urbanized catchments	

orifice. The static pressure may be measured either by the inclined manometer or pressure tranducer devices. Some devices are available with internal PROMs for flow data reduction. In fast flowing water, the dip tube may be protected by a simple still well: a concentrically placed perforated tube. Shortcomings include contaminant build-up on the dip tube in the vicinity of the measuring tube, and relatively low accuracy in the total part of the total pressure range.

Combination Bubbler-Magnetic Devices

These instruments rely on velocity-area measurement to compute instaneous flow rates. Stage measurements are made using conventional transducer bubblers. These data are converted to area measurements using a PROM that describes conduit geometry. Flow velocity measurements are made at the same time with an electromagnetic device located in a band attached to the conduit wall. Using the independent values of area and velocity, the device computes discharge.

Other Monitors

When planning any atmospheric precipitation measurement, contact the National Meteorological Institute or equivalent organization for expert assistance.

Various types of open containers and man-made natural surfaces are used to collect impurities deposited by atmospheric forces. Open containers are generally polyethylene, polypropylene, or glass funnels or cylinders. Various modified gauge types are designed for special purposes.

Precipitation intensity, volume, and duration data should be collected during qualified sample events, and for monitoring programs. Nonrecording gauges are used for measurement by most government hydrological and meteorological services. The ordinary rain gauge used for daily readings is a collector above a funnel leading to a receiver. Continuous registration has also been incorporated into rain gauges. Precipitation recorders in general use are the weighing, tipping-bucket, or float type. If the standard rain gauge is sited in the wind direction, it should be surrounded by an 0.4 m-high screen. A wind shield consisting of a frame of hanging strips is placed within 1 m of the recorder.

QA/QC Measures

A quality assurance/quality control program should be developed and implemented as part of a long-term monitoring program to provide assessment of techniques used during sample collection, storage, and analysis (EPA 1979b, 1980). EPA programs require QA/QC plans to be approved by the EPA prior to sample collection and analysis. The QA/QC plan should specify sample collection and preservation methods, maximum sample holding time, chain-of-custodian procedure, analytical techniques, accuracy and precision checks, detection limits, and data recording and documentation procedures.

SAMPLE STORAGE

The preceding steps will not guarantee a representative sample unless container selection and sample preservation methods meet the required standards. The choice of container and cap materials is very important due to the possibility of interference with constituents to be analyzed. Containers and all elements involved in sampling or compositing operations must be properly cleaned. More detailed information on container types and cleaning is found in EPA 1980b. Recommended operations are as follows:

Container Selection

Containers can introduce positive or negative errors in trace metal and inorganic measurements by contributing contaminants through leaching or surface desorption or, depleting concentrations through absorption. Samples to be analyzed for toxic metals can be stored in 1-l polyethylene or glass bottles with polypropylene caps. Teflon lid liners should be purchased or cut from sheet teflon and inserted in caps to prevent possible contamination from caps supplied with bottles.

Container Cleaning

Due to the sensitivity of tests examining waterborne trace metals, sample containers must be thoroughly cleaned. The following schedule must be followed for the preparation of all sample bottles and accessories, whether glass, polyethylene, polypropylene or Teflon:

- Wash with detergent and tap water
- Rinse with 1:1 nitric acid
- Rinse with tap water
- Rinse with 1:1 hydrochloric acid
- Rinse with tap water
- Triple rinse with distilled (or deionized) water

SAMPLE PRESERVATION

Water samples are susceptible to rapid physical or biological reactions that may take place between sampling and analysis. This time period can exceed 24 hr due to laboratory capacity needed to handle unpredictably varying amounts of samples resulting from aleatory rainfalls (Geiger 1981).

Preservation techniques are recommended to avoid sample changes resulting in large errors. Refrigeration of samples at 4°C is commonly used in fieldwork and helps to stabilize samples by reducing biological and chemical activity. All samples except metals must be refrigerated.

In addition to refrigeration, specific techniques are required for certain parameters. They consist of the addition of chemical compounds, biocides, etc. More detailed information can be found in EPA 1979a, 1980b.

The decision to eliminate a portion of the drainage system from further sampling must include a review of data QA/QC procedures. Review of contaminant data for drainage systems must be performed to ensure that analytical results are properly interpreted, and that detection of potential sources is not missed because of field or laboratory constraints.

Analysis of Pollution Data

When considering the significance of runoff pollutant contributions, both concentrations and total loadings must be examined. Receiving water concentrations are usually of prime concern. In principle, if pollutant concentrations do not exceed certain allowable maxima, detrimental effects will not occur. However, for many pollutants, allowable concentrations in water are not known. Because of sedimentation, accumulation of benthal deposits such as phosphorous, hydrocarbons, and heavy metals may be more significant than concentrations in the water. Receiving water column concentration and benthal accumulation depends more on mass loads of pollutants than of pollutant concentrations in the runoff.

Judging from the above, an estimate of mass loadings of the principal pollutants are of great importance for planning and management purposes. Priority should be placed on obtaining a reliable estimate of mass loadings entering a body of water because of urban runoff during a specific time, such as a year, and data collection plans should be so designed.

STORM LOADS

Data from nonpoint source monitoring studies are usually reduced to a pollutant load per storm basis. An event expected concentration (EMC) is multiplied by the value of runoff. EMCs are calculated by integrating the pollutograph (instantaneous load with time) with the hydrograph. After sampling several storms, these load per storm data are used to estimate annual load from the basin. It is assumed that the monitored storms are representative samples of storms usually occurring in the basin during the year (Whipple 1983).

ANNUAL LOADS

Regressions of total load versus total runoff from a family of storms give a slope in concentration units, which can be used to predict pollutant load for a specific quantity of runoff. To better represent the actual runoff process, base flows were abstracted from storm runoff or low-flow loads from storm load. Good correlations have been found using log-load versus log runoff volume, reflecting log-normal distribution of the concentration data. Nonpoint source data are usually log-normal distributed, as are hydrologic events.

After mass loadings for a given land-use type are accumulated over a considerable period of time, results can be expressed in terms used to estimate loadings from that type of land use for the rest of the watershed(s) of interest. The approaches below are commonly used:

1. Annual loading/area of given land use, lbs/acre/yr
2. Annual loading/curb mi of given land use, lb/mi/yr
3. Annual loading/traffic volume, lbs/vehicle/yr
4. Annual loading/air pollution index, lbs/avg in/yr
5. Annual loading/runoff volume, lbs/million gal
6. Annual loading/precipitation amount, lbs/in (for specific area)

Number 1 assumes that pollution varies according to land use. This is the most commonly used method of predicting loading under future conditions. Number 2 assumes that pollution loading varies with the number of curb mi in various stages of development. Numbers 3 and 4 make similar assumptions regarding automobile traffic and air pollution. Numbers 5 and 6 are designed to convert loading data from specific storm events to annual average loadings, which are then converted to relationships with land use for predictive purposes.

SIMULATION MODEL CALIBRATION

Monitoring data can be used to calibrate sophisticated nonpoint source computer models. These models attempt to interpret the mechanisms involved in nonpoint source generation. Alternatives can then be evaluated using computer-simulated processes. Models exist for different types of basins, levels of complexity, and nonpoint source problems. They need good calibration data.

Statistical Analysis

Urban hydrological phenomena, especially those involving storms, give historical data that can be observed only once, and then will not occur again. Such collected data form an ever-growing sample of measurements. Even if some phenomena can be described by means of physical or rational theory, the input, rainfall, is commonly stochastic in nature and whole phenomena can be amenable to statistical interpretation and probability analysis.

Table 6.4.4 summarizes data analysis methods along with examples and references. Although it is impossible to summarize the many references in general or applied statistics, good basic knowledge is provided in books such as Kendall and Stuart's theory of statistics, or Haan's (1977) statistical methods in hydrology.

—Kent K. Mao
David H.F. Liu

References

Alley, W.M. 1977. *Guide for collection, analysis, and use of urban storm-water data,* conference report. American Society of Civil Engineers (ASCE) New York, N.Y.: American Society of Civil Engineers (ASCE).

Geiger, W.F. 1981. Continuous quality monitoring of storm runoff. *Water Science Technology.* Vol. 13, pp. 117–123. Munich: IAWRR/Pergamon Press Ltd.

Gideometeozdat. 1984. *Complex assessments of surface water quality.* (In Russian). p. 140. Leningrad.

Haan, C.T. 1977. *Statistical methods in hydrology.* Ames, Iowa: The Iowa State University Press.

Shelley, P.E., and G.A. Kirkpatrick. 1975. *An assessment of automatic flow samplers—1975.* U.S. Environmental Protection Agency (EPA). U.S. EPA 600-2-75-065. Washington, D.C.

Smoot, G.F. 1975. *A rain-runoff quantity—quality collection system.* Proceedings of a research conference on Urban Runoff Quantity and Quality. American Society of Civil Engineers (ASCE). New York, N.Y.

U.S. Department of Interior, Bureau of Reclamation. 1975. *Water Measurement Manual.*

U.S. Environmental Protection Agency. 1979a. *Methods for chemical analysis of water and waste.* Washington, D.C.: EPA.

U.S. Environmental Protection Agency. 1979b. *Monitoring requirements, methods, and costs for the nationwide urban runoff program.* EPA Report 600-9-76-014. Washington, D.C.

U.S. Environmental Protection Agency. 1980. *Monitoring toxic pollutants in urban runoff, a guidance manual.* U.S. Environmental Protection Agency (EPA), Office of Water Regulation and Standards.

Wong, J. and J. Marsalek. 1981. *Persistent toxic substances in urban runoff.* Proceedings of Storm Water Management Model Users Group Meeting, Niagara Falls, Ontario, Canada: U.S. Environmental Protection Agency (EPA). EPA Report, pp. 455–468.

Whipple, W., et al. 1983. *Stormwater management in urbanizing areas.* Englewood Cliffs, N.J.: Prentice-Hall.

6.5
DISCHARGE TREATMENT

Three types of treatment are used for wastewater discharges: biological, physical-chemical, and physical processes. Some systems use two or all types to achieve best water quality. The efficiency of various storm water and combined sewer overflow (CSO) treatment processes is given in Table 6.5.1 (Lager et al. 1977).

Biological Processes

The biological processes used for point sources are difficult to implement for stormwater discharges, which have low biochemical oxygen demand (BOD) nonpoint concentrations. These processes perform poorly or not at all when treating flows with irregular quantity or quality. It is very difficult to keep the biota alive between storm events. Wet weather reduces low organic concentrations, and the biomass is sensitive to toxic substances often present in urban stormwater runoff.

Physical-Chemical Processes

Physical-chemical processes show promise in overcoming shock loadings. Chemical coagulants enhance the separation of particles from liquid. Chemical addition is also effective in removing phosphorous, metals, and some organic colloids.

Physical Processes

Successfully demonstrated physical processes include fine-mesh screening, fine-mesh screening/high-rate filtration, sedimentation, sand and peat-sand filtration, fine-mesh screening/dissolved-air flotation, and swirl separation.

SWIRL-FLOW REGULATOR-CONCENTRATOR

The dual purpose swirl-flow regulator-solids-concentrator has shown a potential for simultaneous quality control (Field 1990). These devices have been applied to CSO; however, they can also be used for storm water runoff pollution control.

The swirl concentrator uses a swirl action to separate particles from liquids (Figure 6.5.1). Flow from combined sewers enters a diversion chamber and bar screen, removing the debris. The swirl facility is automatically activated when storm flows enter the lower portion of the circular chamber. Rotary motion causes liquids to follow a long spiral path, to be discharged from the chamber top through a downshaft. This overflow water can be disinfected and discharged or stored for later treatment. Because a flow deflector prevents chamber flow from completing its first revolution and merging with continuing inlet flow, there is a gently swirling rotational movement.

The settleable solids entering the chamber are spread over the full cross-section of the channel and settle quickly. Solids are entrained along the bottom around the chamber and are concentrated at the foul sewer outlet, where they are transported to the treatment plant.

The scum acts as a baffle, keeping floatables outside the overflow weir and preventing these from overflowing into

TABLE 6.5.1   EFFICIENCY OF VARIOUS STORM-WATER AND CSO TREATMENT PROCESSES

Process	Efficiency (%)				
	Suspended Solids	*BOD$_5$*	*COD*	*Total P*	*TKN*[a]
Physical—Chemical					
Sedimentation	20–60	50	34	20	
without chemicals					38
with chemicals	68	68	45		
Vortex separation	40–60	25–60	50–60		
Screening					
microstrainers	50–95	10–50	35	20	30
rotary screens	20–35	1–30	15	12	10
Sand–peat filters[b]	90	90	NA	70	50
Biological[c]					
Contact stabilization	75–95	70–90		50	50
Biodiscs	40–80	40–80	33		
Oxidation ponds	20–57	10–17		22–40	57
Aerated lagoons	92	91			
Facultative lagoons	50	50–90			

Source: Reprinted from J.A. Lager, W.G. Smith, W.G. Lynard, R.M. Finn, and E.J. Finnemore, 1977, *Urban stormwater management and technology: updates and users' guide* (U.S. Environmental Protection Agency (EPA), EPA Report 600–8–77–014, Municipal Environmental Research Laboratory, Office of Research and Development, Washington, D.C.).
[a]Total Kjeldahl nitrogen.
[b]After Galli (1990), peat-sand filters are similar to biological anaerobic-aerobic slow filters. They are applicable for treatment of urban runoff.
[c]Biological treatment is feasible only for CSOs.

the clean effluent. Floatables are directed by a floatable deflector to a floatable trap. The floatable trap is connected to a floatable storage area under the clear overflow weir plate. Floating material is drawn beneath the weir plate by the vortex and dispersed around the downshaft. Floating solids are retained here until after a storm event, when the water level recedes in the swirl chamber. As this occurs, trapped floatables are dropped and enter the foul sewer outlet, where they are transported to a sewage treatment plant.

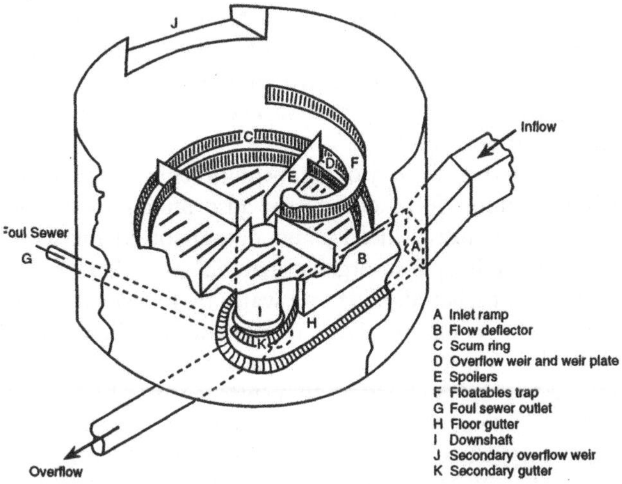

A partial list of U.S. installations with experience in swirl-flow regulator-concentrator use was presented by Pisano (1989).

SAND FILTERS

Sand systems are usually off-line. A typical sand filtration system is comprised of an inlet structure with a presetting basin, a flow disperser, filtration media, an underdrain system, and a basin liner (Figure 6.5.2). For piped storm water systems, the inlet structure might be a manhole using a weir to divert low flows into the filtration system. For open-channel conveyance systems, the inlet might be a weir constructed within the flow path to divert low flows to the filtration system, while allowing higher flows to bypass the filtration system. Without a presettling basin, the filter medium may quickly become plugged with large sediments. This basin may not be necessary, if the sand filtration basin is used in place of an API oil/water separator, and if the contributing drainage area is small and completely impervious.

The primary pollutant removal mechanisms are filtration and sedimentation. Particulate matter such as sediments, oils and greases, and trace metals are removed by filtration as stormwater percolates through the sand filter. Sedimentation removes large particles, and filtration removes silt and clay-size particles.

Over time, sediment eventually penetrates the filter media surface, requiring replacement of the filter media.

A   Inlet ramp
B   Flow deflector
C   Scum ring
D   Overflow weir and weir plate
E   Spoilers
F   Floatables trap
G   Foul sewer outlet
H   Floor gutter
I   Downshaft
J   Secondary overflow weir
K   Secondary gutter

FIG. 6.5.1   An isometric view of a swirl regulator-concentrator.

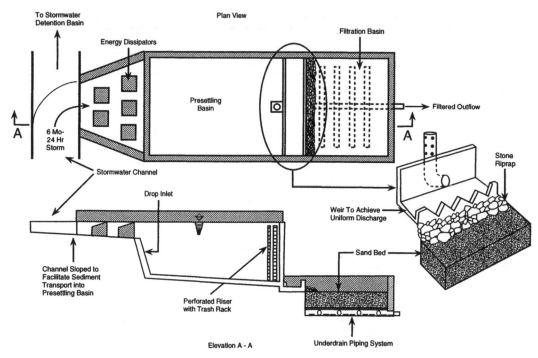

FIG. 6.5.2 Conceptual sand filtration basin system. (Reprinted from the City of Austin, 1988.)

Maintenance requirements can be intensive, depending upon sediment concentrations in surface runoff. Fifty acres is recommended as the maximum contributing drainage area for a sand filtration system (Schueler, Kumble and Heraty 1992).

ENHANCED FILTERS

Enhanced (peat-sand) filters use layers of peat, lime, and/or topsoil, and may also use a grass cover (Figure 6.5.3) to remove particulate pollutants. To minimize clogging, both sand and enhanced filters should be preceded by a solid-removing unit, such as a pond or a filter strip.

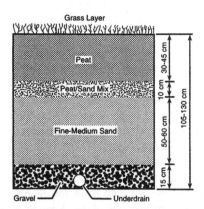

FIG. 6.5.3 Peat sand filter for storm water treatment. (Reprinted from J. Galli, 1990, *Peat sand filters: A proposed storm water management practice for urban areas* [Department of Environmental Programs, Metropolitan Washington Council of Governments, Washington, D.C.].)

Peat-sand filters provide high phosphorous, BAD, nitrogen and silt removal. Peat has a high removal affinity for adsorbing and removing toxic compounds (Novotny 1994), hence peat-containing filters are effective for removing priority pollutants.

COMPOST FILTERS

W&H Pacific conceived the idea of utilizing yard debris compost as a treatment and filtration medium for stormwater runoff. This medium removes organic and inorganic pollutants through adsorption, filtration, and biological processes (ion exchange and bioremediation). The Compost Storm Water Treatment System (CSF℠) has been constructed at eight different sites throughout Oregon. Six of the eight systems are enclosed facilities, located below grade, while the remaining two are open channel systems retrofitted into existing swales. The technology is being tested and field modified.

The filtering capacity of the medium removes sediments from the runoff. Ion exchange and adsorption removes oils and greases, heavy metals, and non-dissolved nutrients. Following adsorption, organic material is further broken down into carbon dioxide and water by microbial action within the compost. Treated stormwater then passes through a 6 in to 8 in gravel layer underneath the filtering media, and is conveyed to a surface water body or to a storm drainage system by an underdrain system.

Prototype test results for nine events show good solids removal: 67% removal of COD, 40% removal of total phosphorous, 67% removal of copper, and better than

87% removal of zinc, aluminum and iron. The leaf compost has very good cation exchange capacity. However, like sand and enhanced filters, operating life depends on the frequency of preventive maintenance.

—Kent K. Mao

References

Field, R. 1990. Combined sewer overflows: Control and treatment. In *Control and treatment of combined-sewer overflows.* Moffa, P.E. ed. New York, N.Y.: Van Nostrand Reinhold.

Lager, J.A., W.G. Smith, W.G. Lynard, R.M. Finn, and E.J. Finnemore. 1977. *Urban stormwater management and technology: Updates and users' guide.* U.S. Environmental Protection Agency (EPA). EPA Report 600-8-77-014. Municipal Environmental Research Laboratory, Office of Research and Development.

Novotny, V., and H. Olem. 1994. *Water quality: Prevention, identification, and management of diffuse pollution.* New York, N.Y.: Van Nostrand Reinhold.

Pisano, W.C. 1989. Recent United States experience with designs and new German technology. In *Design of urban runoff quality controls.* L.A. Roesner, B. Urbonas, and M.B. Sonnen, eds. American Society of Civil Engineers (ASCE). New York, N.Y.

Schueler, T.R., P.A. Kumble, and M.A. Heraty. 1992. A current assessment of urban best management practices. Techniques for reducing non-point source pollution in the coastal zone. *Technical guidance manual.* Metropolitan Washington Council of Government. Office of Wetlands, Oceans, and Watersheds. Washington, D.C.

Bibliography

Aravin, V.L., and S.N. Numerov. 1965. *Theory of fluid flow in undeformable porous media.* New York: Daniel Davey.

Bear, J. 1972. *Dynamics of fluids in porous media.* Elsevier, Amsterdam.

Beasley, D.B. 1976. Simulation of the environmental impact of land use on water quality. In *Best management practices for non-point source pollution control.* U.S. Environmental Protection Agency (EPA). EPA Report 905-9-76-005. Washington, D.C.

Bennett, G.D. 1976. *Introduction to groundwater hydraulics,* Techniques of Water Resources Investigations, Chap. B2, Book 3, U.S. Geological Survey, Washington, D.C.

Bowen, R. 1980. *Groundwater.* Barking, Essex, England: Applied Science Publishers Ltd.

Cooper, H.H. 1966. The equation of groundwater flow in fixed and deforming coordinates. *J. Geophys. Res.* 71, no. 20:4785-4790.

Council on Environmental Quality. 1980. *Environmental quality—1978: The ninth annual report of the council of environmental quality.* Washington, D.C.: U.S. Government Printing Office.

Crawford, N.H., and R.K. Linsley. 1966. Digital simulation in hydrology: The Stanford Model IV. *Technical Report No. 39.* Stanford University, Department of Civil Engineering. Palo Alto, Calif.

Davis, S.N., and R.J.M. DeWiest. 1966. *Hydrogeology.* New York: John Wiley & Sons, Inc.

DeVries, J.J. 1975. *Groundwater hydraulics.* Aqua-Vu, Ser. A, no. 6, Communications of the Institute of Earth Sciences. Amsterdam: Free Reformed University.

DeWiest, R.J.M. 1965. *Geohydrology.* New York: John Wiley & Sons, Inc.

DeWiest, R.J.M., ed. 1969. *Flow through porous media.* New York: Academic Press, Inc.

Hantush, M.S. 1964. Hydraulics of wells. In *Advances in hydroscience.* Vol. 1, edited by V.T. Chow. New York: Academic Press, Inc.

Harr, M.E. 1962. *Groundwater and seepage.* New York: McGraw-Hill Book Company.

Heaney, J.P., and W.C. Huber. 1972. *Storm water management model; refinements, testing and decision making.* University of Florida, Department of Environmental Science. Gainesville, Fla.

Heath, R.C. 1983. *Basic groundwater hydrology.* Water Supply Paper 2220, U.S. Geological Survey. Washington, D.C.

Hubbert, M.K. 1940. The theory of groundwater motion. *J. Geol.* 48:785-944.

Jacob, C.E. Flow of groundwater. In *Engineering hydraulics,* edited by H. Rouse. John Wiley.

Javandel, I., C. Doughty, and C.F. Tsang. 1984. *Groundwater transport: Handbook of mathematical models.* Washington, D.C.: American Geophysical Union.

Lohman, S.W. 1972. *Groundwater hydraulics.* Professional paper 708, U.S. Geological Survey. Washington, D.C.

Marino, M.A., and J.N. Luthin. 1982. *Seepage and groundwater.* Developments in Water Science Series no. 13. New York: Elsevier Science Publishing Co., Inc.

McWhorter, D.B., and D.K. Sunada. 1977. *Groundwater hydrology and hydraulics.* Fort Collins, Colo.: Water Resources Publications.

Meinzer, O.E. [1923] 1960. *Outline of groundwater hydrology.* Water Supply Paper 494, U.S. Geological Survey. Washington, D.C.

Novotny, V., and G. Chesters. 1981. *Handbook of nonpoint pollution: Sources and management.* New York, N.Y.: Van Nostrand Reinhold.

Novotny, V., R. Imhoff, M. Olthoff, and P.A. Krenkel. 1989. *Karl Imhoff's handbook of urban drainage and wastewater disposal.* John Wiley.

Novotny, V., and H. Olem. 1994. *Water quality: Prevention, identification, and management of diffuse pollution.* New York, N.Y.: Van Nostrand Reinhold.

Petersen, D.F. 1957. Hydraulics of wells. *Trans. Am. Soc. Cir. Eng.* 122:502-517.

Pinneker, E.V., ed. 1983. *General hydrogeology.* Cambridge: Cambridge University Press.

Polubarinova-Kochina, P.Y. 1962. *Theory of groundwater movement.* Translated from Russian by R.J.M. DeWiest. Princeton, N.J.: Princeton University Press.

Todd, D.K. 1964. Groundwater. In *Handbook of applied hydrology,* edited by V.T. Chow. New York: McGraw-Hill Book Company.

United Nations Educational, Scientific and Cultural Organization (UNESCO). 1987. Manual on drainage in urbanized areas. *Vol. 1, Planning and design of drainage systems.* Paris, France: UNESCO Press.

U.S. Bureau of Reclamation. 1960. *Studies of groundwater movement.* Technical Memorandum 657. Denver, Colo.: U.S. Dept. of Interior.

U.S. Department of Interior, Water and Water Resources Service. [1977] 1981. *Groundwater manual.* Washington, D.C.: U.S. Government Printing Office.

U.S. Soil Conservation Service (SCS). 1975. Procedure for computing street and rill erosion on project areas. *SCS technical release no. 51.* Washington, D.C.

U.S. Soil Conservation Service (SCS). 1975. Urban hydrology for small watersheds. *SCS technical release no. 55.* Washington, D.C.

Whipple, W., Jr. et al. 1983. *Storm water management in urbanizing areas.* Englewood Cliffs, N.J.: Prentice Hall.

Index